International Series in Operations Research & Management Science

Volume 227

More information about this series at http://www.springer.com/series/6161

Geert-Jan van Houtum • Bram Kranenburg

Spare Parts Inventory Control under System Availability Constraints

Springer

Geert-Jan van Houtum
Eindhoven University of Technology
Eindhoven, The Netherlands

Bram Kranenburg
Consultants in Quantitative
Methods CQM B.V.
Eindhoven, The Netherlands

ISSN 0884-8289 ISSN 2214-7934 (electronic)
International Series in Operations Research & Management Science
ISBN 978-1-4899-7917-9 ISBN 978-1-4899-7609-3 (eBook)
DOI 10.1007/978-1-4899-7609-3

Springer New York Heidelberg Dordrecht London

Softcover reprint of the hardcover 1st edition 2015

Printed on acid-free paper

Springer Science+Business Media LLC New York is part of Springer Science+Business Media (www.springer.com)

To Astrid, Laura, and Milou

To Marleen, Else, and Leonie

Preface

When looking back, we can say that the journey leading to this book started in the fall of 1994. At that time, the first author started to study the spare parts management at the Royal Netherlands Navy (RNN), which was part of a large maintenance project of the RNN and the University of Twente. Professor Henk Zijm suggested to use the METRIC type models of Craig Sherbrooke as described in his book of 1992. Shortly later, a Navy Officer, Jan Willem Rustenburg, started his PhD project on this topic, with Henk Zijm and the first author as supervisors. In this project, the METRIC type models were applied to cases at the RNN, an exact evaluation method for a general multi-echelon, multi-indenture system was developed, and problems with yearly budget constraints were studied. The cases showed the strength of the so-called system approach in comparison to a straightforward item approach.

In 1999, the first author moved to Eindhoven University of Technology, and he started to collaborate with multiple Original Equipment Manufacturers (OEMs) that manage large spare parts networks to serve their customers. Over the years, many master and PhD projects were carried out with companies such as ASML, DAF, IBM, KLM, Marel, NedTrain, Océ, Philips Healthcare, and Vanderlande Industries. In the spare parts networks of most of these companies, it is not common to backorder demands when stockouts occur, as assumed in the METRIC type models. Instead, lateral transshipments and emergency shipments are applied in order to satisfy a demand as quickly as possible. This avoids long and costly downtimes of the manufacturing and transportation equipment that is supported by the spare parts networks. Further, it was important to deal with multiple customer classes, each with their own target system availability level. This motivated the start of the final project of the second author in 2002 for the postgraduate program Mathematics for Industry. In this project, we collaborated with ASML, an OEM of expensive lithography equipment for the semiconductor industry. Subsequently, this project was continued with a PhD project.

In his PhD project, the second author developed new multi-item spare parts models and followed a system approach for all of them. Aspects that were included were commonality, service differentiation, lateral transshipments, and emergency shipments. The people at ASML, Harrie de Haas, Eric Messelaar, and Harold Bol, were directly involved in the modeling phase. The potential of these new models for the practice of ASML was determined via case studies. One of these case studies showed a huge potential. That was for the multi-item, multi-location model with lateral transshipments. ASML used already lateral transshipments at the operational level, but this was not included in the spare parts planning model at the tactical planning level. The case study showed that including lateral transshipments at the tactical planning level gave an improvement potential of 30 %; see Chap. 5 for more details. This had an enormous impact. ASML implemented the lateral transshipments model in 2005, and they use the model for their tactical planning since early 2006, i.e., for already more than 9 years.

The PhD dissertation of the second author achieved both a high scientific level and a high practical relevance. It received a great appreciation from both the academic and business world. On top of that, the second author received the EURO Doctoral Dissertation Award 2007 of the Association of European Operational Research Societies for the best PhD dissertation defended at a European university within the Operations Research area. At that time, the idea arose to write the underlying book.

This book focuses on the tactical planning level for spare parts management. It describes a series of multi-item inventory models and presents exact and heuristic optimization methods, including greedy heuristics that work well for real-life size problems. The intended audience consists of graduate students, starting scholars in the field of spare parts inventory control, and spare parts planning specialists in the industry.

After an introductory chapter, we present a basic, single-location model in Chap. 2. This chapter forms a basis for all other chapters. Next, Chaps. 3 and 4 are on single-location models with multiple machine types and/or machine groups. In these chapters, we deal with commonality and service differentiation, respectively. In Chap. 5, we present the multi-location model with lateral transshipments. This chapter contains the successful case as mentioned above. In Chaps. 6 and 7, we deal with the classical METRIC model and its generalization to multi-indenture systems, respectively. Finally, in Chap. 8, we go back to a single-location model, and we have an explicit modeling of the repair capacity for failed parts and the priorities that one can set there. Chapters 3–5 are based on joint papers of the authors. Chapter 2 is based on lecture notes, written together with Kristel Hoen. Chapters 6–8 are based on papers with multiple other authors: Hartanto Wong, Dirk Cattrysse, Jan Willem Rustenburg, Henk Zijm, Ivo Adan, and Andrei Sleptchenko.

Various chapters of the book have been used in a master's course at Eindhoven University of Technology and in a PhD course of the Dutch Network on the Mathematics of Operations Research. The required pre-knowledge consists of probability theory and basic knowledge on Markov processes and queueing theory. We mainly used Chaps. 1, 2, 5, and 6 in these courses. Generally, we advise to start with Chaps. 1 and 2, and after that, one can do each of the Chaps. 3–6 and 8. Chapter 7 builds on Chap. 6, and thus Chap. 7 can only be treated after Chap. 6 has been studied.

We would like to thank many people, without whom this book would not have been there. First, we thank all co-authors as mentioned above. Next, we want to thank a number of colleagues at Eindhoven University of Technology. We are very grateful to Engin Topan, who executed the computational experiment in Sects. 3.4.1 and 3.4.2, gave a lot of feedback on all chapters, and helped us with the index at the end of the book. We also thank Joachim Arts, Erwin van Wingerden, and Martijn van Aspert for their feedback. We are grateful to Eindhoven University of Technology for its support during the whole process. Further, we want to thank all involved people at ASML for their collaboration, since we started to work with them. We also thank all involved people at Springer for their great support, regarding all kinds of things that have to be arranged when writing a book. In particular, we like to thank Fred Hillier, who as an editor kept us asking about the progress at a regular basis and who had always stimulating words when we reported that the progress was less than planned. Finally, we thank the many colleagues, friends, and family members who showed their interest in our book and denoted that they would like to see or even read the book when finished. Well, go ahead, our book is available now!

Eindhoven, The Netherlands
February 2015

Geert-Jan van Houtum
Bram Kranenburg

Contents

About the Authors

Geert-Jan van Houtum is Professor of Maintenance and Reliability at Eindhoven University of Technology since 2008. Prior to that, he filled positions as assistant/associate professor at the University of Twente (1994–1998) and Eindhoven University of Technology (1999–2007) and as visiting professor at Carnegie Mellon University (2001). He obtained his MSc and PhD degrees in Applied Mathematics from Eindhoven University of Technology in 1990 and 1995, respectively. He does research on the maintenance and reliability of capital goods and, in particular, on: (i) design and control of service supply chains, (ii) maintenance concepts, and (iii) design for availability. He publishes in journals such as Operations Research, Manufacturing & Service Operations Management, European Journal of Operational Research, and International Journal of Production Economics. He is associate editor of Manufacturing & Service Operations Management, OR Spectrum, the Flexible Services and Manufacturing Journal, and Mathematical Methods of Operations Research. For much of his research, he collaborates with companies such as ASML, Gordian, IBM, NedTrain, Océ, Philips, Marel, and Vanderlande Industries. He is scientific director of the Beta Research School for Operations Management and Logistics.

Bram Kranenburg is Senior Consultant at *Consultants in Quantitative Methods CQM* in Eindhoven. He works at CQM since 2006. The projects he is involved in concern supply chain analytics, modeling, innovation, and optimization projects. He obtained his MSc degree in Industrial Engineering and Management at the University of Twente in 2000, his MTD (Master of Technological Design) degree in Mathematics for Industry at Eindhoven University of Technology in 2003, and his PhD degree at Eindhoven University of Technology, Department of Technology Management, in 2006. Geert-Jan van Houtum and Ton de Kok were the supervisors of the PhD project. Both the final project of Mathematics for Industry and the PhD project have been carried out in collaboration with ASML. For his PhD dissertation, Bram received the EURO Doctoral Dissertation Award 2007 of the Association of European Operational Research Societies for the best PhD dissertation defended at a European university within the Operations Research area.

Chapter 1
Introduction

1.1 Importance of Capital Goods, Maintenance and Spare Parts

This book deals with inventory control of spare parts for capital goods. Capital goods are machines or products that are used by manufacturers to produce their end-products or that are used by service organizations to deliver their services. For example:

- Lithography machines are used by manufacturers of semiconductors;
- Medical equipment is used by hospitals to diagnose and cure patients;
- Trains are used by a service organization such as Dutch Railways to transport customers to their desired destinations;
- Baggage handling systems are used by airports to move baggage from passengers to airplanes and vice versa.

Capital goods require maintenance. In general, maintenance may be defined as: "the combination of all technical and associated administrative actions intended to retain an item in, or to restore it to, a state in which it can perform its required function" (cf. the British Standard Institute). Capital goods are used in primary processes of their users, and failure of a capital good may cause standstill of a complete factory, or, in the case of public transportation, can severely impact a large part of a country.

Capital goods have a life cycle consisting of 5 phases; see Fig. 1.1 (cf. Dinesh Kumar et al. [5]). In the first phase, needs and requirements are defined based on feedback from the market and knowledge of technical possibilities. Next, the system is completely designed. After that, multiple units of the system are produced. Then, in the exploitation phase, the systems are used, generally for long periods (say 10–40 years). Finally, the system is disposed of. The manufacturer of a capital good, also denoted as Original Equipment Manufacturer (OEM), is generally the leading party in the first three phases. The role of the OEM in the last two phases may vary from a very limited to a leading role.

G.J. van Houtum, B. Kranenburg, *Spare Parts Inventory Control under System Availability Constraints*, International Series in Operations Research & Management Science 227, DOI 10.1007/978-1-4899-7609-3_1

Fig. 1.1 Life cycle of a capital good

What matters for the users of capital goods are the costs during the entire life cycle. These are known as the Total Cost of Ownership (TCO). The costs during the first three phases of the life cycle are reflected in the sales price for new systems. We call these costs the acquisition costs, which in many cases are high. The price of a lithography machine as needed for the production of semiconductors is 20–50 million Euros, an MRI scanner costs 1–2 million Euros, a baggage handling system at a major airport may range up to 300–400 million Euros, and one Joint Strike Fighter will have a price of 80–100 million Euros.

The rest of the TCO occurs after the purchase of the new system. During the exploitation phase, costs of multiple types arise, with maintenance and downtime accounting for the largest proportion. Maintenance costs consist of all the resources needed for maintenance, which may be executed by the user itself or by the OEM or a third party. In any case, these costs include the costs of spare parts, service/maintenance engineers, infrastructure and management. Downtime costs may involve direct costs, such as those caused by a reduction in the output of a factory, and indirect costs, such as those caused by loss of reputation and resulting loss of future revenues. Finally, in the disposal phase, there can be some disposal costs. These costs may be significant if systems contain environmentally unfriendly materials. In many cases, the disposal costs are low. In some cases, systems or parts of systems may be refurbished and can be re-used, so that disposal may even lead to a revenue instead of a cost.

To give an impression of how high the costs of a capital good may be after purchase, Fig. 1.2 shows how the TCO of an engineer-to-order system is divided over the acquisition, maintenance and downtime costs (cf. Öner et al. [10]). In this example, costs were determined for one particular system with an assumed lifetime of 20 years, and all costs were discounted (with a discount rate of 5 % per year).

As we can see in Fig. 1.2, the acquisition costs constitute only one-third of the TCO, while the rest of the TCO consists of maintenance and downtime costs. For other systems, we may get different numbers, but generally the message is the same: the acquisition costs account for only a fraction of the TCO. So, a buyer of a new system implicitly makes an investment that is 2–4 times as much as the acquisition costs alone.

The maintenance of capital goods is a major industry in itself, which is shown by different types of statistics. According to a study by Aberdeen Group [1], global spending on after-sales services totaled more than 1,500 billion US dollars annually; and spare parts sales and services in the United States accounted for 8 % of the annual Gross Domestic Product. Deloitte [4] carried out a study among a group of more than 120 large OEM's in Europe, North America and the Asia-Pacific region with combined revenues of more than 1,500 billion US dollars. They reported that 26 % of these revenues (390 billion US dollars) came from services. Another inter-

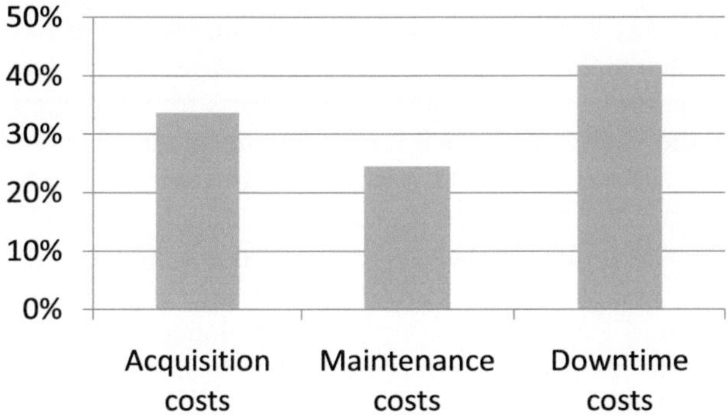

Fig. 1.2 Division of TCO for an engineer-to-order system

esting statistic was generated by Lam [7]. He looked at the total costs of commercial airlines, and found that maintenance accounts for around 10 % of the airlines' total costs, which is about the same as fuel and travel agents' commission.

Spare parts play an important role in maintenance. Capital goods require preventive and corrective maintenance. Preventive maintenance is scheduled in advance, and may require spare parts. Corrective maintenance is carried out upon failure of a system. Typically, if the failure is caused by a failing part, then the corrective maintenance is done on a repair-by-replacement basis: The failed part is removed from the machine and replaced by a new or as good as new spare part. This avoids a too long and costly downtime. For this corrective maintenance, a repairman is needed as well as the proper spare part. Since the demand for spare parts is not known in advance, companies that provide the after-sales service have to keep spare parts inventory. Flint [6] stated that, already in 1995, the world's spare parts inventory in the aviation industry amounted to 45 billion US dollars. Any means to downsize this stock, without decreasing customer service, would be more than welcomed by the aviation industry. Also in other industries, large amounts of money are invested in spare parts inventory and this has increased over the years. Because of these large amounts of money involved, there is a great interest in more efficient methods for spare parts management. Notice that smart approaches for spare parts provisioning reduce both the maintenance costs and downtime costs.

1.2 Trends

With respect to the maintenance of advanced capital goods, we see a number of trends that reinforce each other. First, systems are becoming more and more advanced, with increasing numbers of components, which means their maintenance is also becoming increasingly complex, especially for buyers/users having only limited numbers of systems. Second, one sees that, even though the number of components has increased, the reliability of systems has increased as well or stayed at the same level. This is so because the reliability of components has increased severely over time. Although this is positive from a reliability point of view, it complicates spare parts management. One may need spare parts for many more different components, but at the level of individual users the demand rate per component is very low, so low that even statistical analysis is hardly possible. Third, the buyers/users of capital goods demand ever-higher system availability levels, which implies that the whole maintenance function has to be executed at a higher level. Fourth, nowadays buyers/users look explicitly at the TCO when they buy new systems, or equivalently at the cost of a capital good per produced product or the cost per serviced unit. In other words, they want both a high system availability and a low TCO (where TCO now excludes downtime costs).

To deal with the above trends, we can observe technological innovations and innovations in business models. In the area of business models, we see that more and more users outsource maintenance activities to either the OEM's or third parties (see also Oliva and Kallenberg [9] and Cohen et al. [3]). This may start with service contracts for spare parts. Then the OEM or the third party can enable that spare parts are shared by multiple users in a certain way, and, by getting demand data for many systems, they can still apply statistical analysis. Later on also the activities by service engineers may be outsourced, and in the end one may have full service contracts, under which the service provider (OEM or third party) is fully responsible for all maintenance. The buyer/user then simply pays a fixed service cost per year, and it becomes the service provider's task to manage everything. In the most extreme case the buyer/user does not even buy the system, but just buys the function of the system.

Technological innovations start mainly in the design phase of technical systems and are as follows. To meet the pressure on short downtimes after failures, we see OEM's who make their systems more modular, with well-structured repair processes and simple repair-by-replacement actions. We also see that redundancy is built in. Further, for many technical systems, it is nowadays possible to measure parameters of critical components that reveal the degradation behavior. This is done via the process control subsystem and enables the prevention of failures, by executing a preventive replacement at the moment that a certain threshold is exceeded (condition-based maintenance). The power of these measurements is strengthened by developments in the area of communication technology. It is nowadays possible to collect measurements at a central place for larger amounts of systems, which facilitates statistical analysis for the purpose of predictive maintenance. Obviously, the technological and business innovations reinforce each other.

The above trend from a product-oriented to a function-oriented market (or industry) can be clearly seen for many types of capital goods. The speed of this trend and the current position of each market depends strongly on the technological complexity of the equipment, the size of the installed base per user, the importance of high system availability, and the importance of a low TCO. Examples of capital goods for which the market is strongly function-oriented are aircraft engines, lithography machines, and large-scale computer systems.

1.3 Typical Features of Spare Parts Inventories

As stated above, spare parts have to be kept on stock to facilitate preventive and corrective maintenance. If an OEM closes full service contracts with most customers, then he will setup and manage a spare parts network. Often systems are installed at different places in a continent or in the whole world, and then the OEM has a network consisting of one (or more) central warehouse(s) and many local warehouses; see Fig. 1.3. On the righthand side of this figure, we see technical systems which are installed at their users. These systems constitute the so-called *installed base*. A spare parts demand occurs at a technical system and, if possible, then the demand is fulfilled from the local warehouse to which the system is assigned. If the local warehouse is out-of-stock, then either the demand is backordered and fulfilled as soon as a part becomes available in the local warehouse or the demand is satisfied from another local warehouse by a so-called *lateral transshipment* or directly from the central warehouse (or another upstream source) by a so-called *emergency shipment*. The stocks in the local warehouses are replenishment from the central warehouse. The central warehouse is supplied with new parts from external suppliers or internal production departments, and, for repairable items, with ready-for-use parts from internal and external repair shops where failed parts are repaired.

If full service contracts are closed with a third party, then one may get a similar network but then managed by the third party. The same type of network is also obtained when a user of a fleet of systems (such as trains or military equipment) manages the entire maintenance process itself and has its own spare parts network. In all cases, the same principles apply.

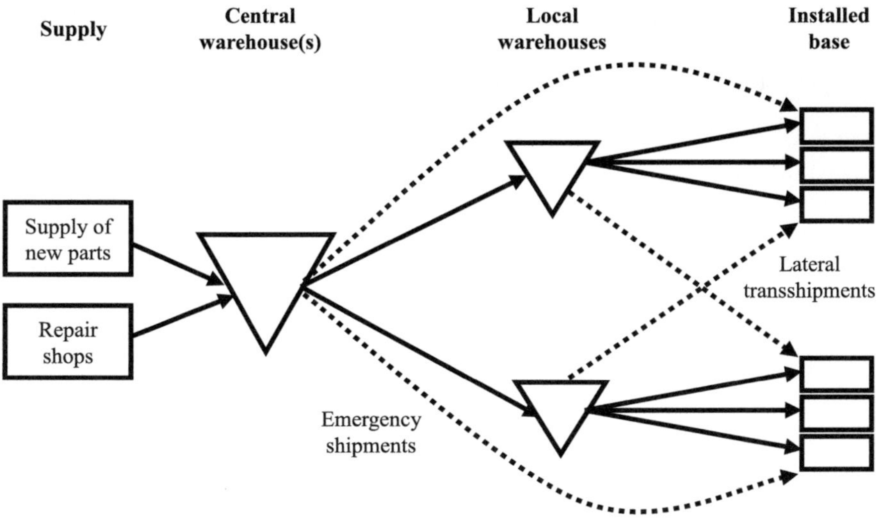

Fig. 1.3 Graphical representation of a typical spare parts network

The main function of a spare parts network as depicted in Fig. 1.3 is to deliver spare parts within specified time constraints. For such a network, two important principles have to be followed for design and planning decisions:

1. *Availability/downtime constraints are formulated at the level of capital goods. Avoid decompositions into lower-level constraints and integrate decisions as far as possible.* In practice, it is common that constraints for downtime due to maintenance are translated into separate constraints for spare parts, or that constraints at a machine level are translated into constraints for single items. While this may to some extent be necessary to manage the whole network or to allow optimization problems to be solved, it is in general better to avoid these decompositions as far as possible.
2. *Create as much pooling effect as possible for the spare parts.* Spare parts exist to meet demands. The demand processes are stochastic (volatile), and buffers need to be created in the spare parts stocks to meet demand peaks. The more demands can be bundled before they are matched with spare parts, the less buffering is needed. In fact, this is something that applies to all supply networks. But special approaches are needed in spare parts networks because of the low demand rates and the tight downtime constraints.

The first principle facilitates that a so-called *system approach* can be followed for the inventory control of spare parts (one also refers to the system approach as a *multi-item approach*). The alternative is a straightforward *item approach*, under which inventories of single items are optimized under decomposed constraints. A system approach allows that one keeps relatively low stocks for expensive items and relatively high stocks for cheap items. This may lead to the same performance in terms of system availability or total downtime, while the total stock investment for the spare parts in Euros is significantly reduced. The system approach has been introduced by Sherbrooke [13, 14] and has led to a large stream of research; see the survey in [2]. Also, this approach has been applied for many real-life problems (see Sect. 2 of [2]), and multiple authors have shown that a system approach can easily result in savings up to 50 % compared to the standard item approach; see e.g. Sherbrooke [14], Rustenburg et al. [11, 12], Thonemann et al. [15], and Wong et al. [16].

Following the system approach means that spare parts planning problems are to be solved via multi-item inventory models. In this book, we will develop such models and we derive heuristics that are effective and efficient for real-life problem instances.

The second principle, creating as much pooling effect as possible, can be applied in various ways. A first way is via the use of lateral transshipments and emergency shipments as depicted in Fig. 1.3. Tight availability or downtime constraints force that, in addition to a central stock, spare parts have to be stocked at short distances from the installed machines. At the same time failure rates are low, which means we have low to very low demand rates at the local warehouses. Excessively high spare parts stocks at the local stock points can be avoided by the use of lateral transshipments and emergency shipments. By those shipments, the whole network will behave as if it is one large, virtual stockpoint. In case lateral transshipments are not applicable between all combinations of local warehouses, but only between local warehouses in the same continent (or region), then we at least get the effect that local warehouses in the same continent start to behave as one large, virtual stockpoint.

Similarly, one can create a pooling effect between spare parts stocks for different machine types with common components. Instead of keeping separate spare parts stocks per machine type, one can create joint spare parts stocks and set availability constraints per machine type. The same idea also applies if one has full service contracts with different service level targets for different groups of customers. It is then preferred to have joint spare parts stocks instead of having separate stocks for different groups, but of course a certain mechanism is needed to get differentiated service levels.

Creating the above pooling effects leads to more complicated control structures and the price of that complexity has to be traded off against the efficiency improvement. With the models developed in this book, we are able to quantify the efficiency improvement.

1.4 Objectives of This Book

In this book, we focus on the tactical planning problems for spare parts networks as described in Sect. 1.3. We describe a series of clean multi-item inventory models for spare parts. We present exact and heuristic methods for the optimization of spare parts stocks, and this includes methods that work for real-life size problems. In each of the models, we assume a class of appropriate policies, being typically base stock policies, because they are easy to use in practice and this enables that real-life size problems can be solved. We see the clean inventory models as building blocks that one can use to construct a planning concept for a spare parts network of a particular company. Notice, that real-life networks are often complex and that they can vary with respect to many features. Therefore it makes sense to construct planning concepts from generic building blocks.

Via the various inventory models that we describe, we will also be able to answer questions such as:

- Should we make one joint planning for multiple machine types with common components, or is it equally good to make a planning per machine type?
- What are the costs and benefits of using lateral transshipments?
- What are the costs per customer class when we have three classes with system availabilities of 98 %, 99 %, and 99.5 %, respectively?
- Or, what is the effect on total costs when a few local warehouses of a network are duplicated in areas with growing numbers of customers?

In other words, by comparing scenarios, we can also support aspects that play a role at a more strategic decision level.

The models that we present are meant for the planning in stationary situations. We assume stochastic demand processes and look at the performance in steady state. This implies that the models can be used for the inventory planning during the exploitation phase, as long as regular procurement and repair of parts is possible. In addition, the models can be used for the initial supply of spare parts when new machines are introduced and installed. Further, the presented models are meant for so-called *critical components*, which have a direct and significant effect on the system performance when they fail. Non-critical components can be planned via straightforward inventory planning methods. One may choose to use such straightforward methods also for the cheap critical parts, for which then high service levels are required. That reduces the numbers of items for which the multi-item models are applied, and thus also the requirements with respect to computational efficiencies.

The main audience for this book is constituted by graduate students of industrial engineering and closely related programs. Much of the material is also appropriate for practitioners, for whom model descriptions, algorithms, and insights of computational experiments are well accessible.

One may notice that the focus on spare parts inventory models for tactical planning problems is the same as for the books of Sherbrooke [14] and Muckstadt [8]. However, for many of the models, there are differences in the main assumptions and the solution approaches. In particular, in this book, we have a much stronger focus on models with emergency and lateral shipments.

1.5 Outline

The rest of this book consist of four blocks. The first block consists of Chap. 2, in which a basic multi-item, single-location inventory model is presented. In this chapter, we consider various system-oriented service measures, we deal with cases without and with emergency shipments, and we discuss multiple methods for the optimization of the spare parts stock. This basic model is a building block that may fit for both the central warehouse or a single local warehouse.

The second block consists of Chaps. 3–5, where building blocks for the planning of local warehouses are discussed. In these models, we assume the use of emergency and lateral shipments. In Chap. 3, we consider a situation with a single stockpoint and multiple machine types with common components. Each machine type has its own system-oriented service target, and we show how to deal with these multiple targets. In this chapter, we will see the benefits of making a joint planning for multiple machine types as a function of the level of commonality. In Chap. 4, we consider a single stockpoint that serves multiple customer classes, each with their own service target. We introduce critical level policies to differentiate between these customer classes, and discuss heuristic procedures for the critical and basestock levels. We show the benefits of the use of the critical levels as a function of sizes of the customer classes and the differences in the service targets. In Chap. 5, we present a model for multiple local warehouses with a so-called partial pooling structure. We distinguish main and regular local warehouses and allow lateral transshipments between pairs of two main local warehouses and from each main local warehouse to each regular local warehouse. We present an approximate evaluation procedure and a heuristic procedure for finding near-optimal basestock policies. We show the benefits of using lateral transshipments (i.e., of explicitly incorporating their use in a decision support model for the tactical planning level), and how many local warehouses have to be defined as main local warehouse in order to get most of the maximum benefits of using lateral transshipments. In this chapter, we also discuss case results at ASML, an OEM in the semiconductor supplier industry, and we report on the successful implementation of this model at ASML.

The third block is constituted by Chaps. 6 and 7, where we consider multiple echelons and a so-called multi-indenture structure. To repair a repairable part, one often needs parts at a lower level in the material breakdown structure of a technical system. When both the inventories of the repairables and the lower level parts are considered in one model, we call such a model a multi-indenture model. In Chap. 6, we study a multi-item, two-echelon, single-indenture model with a service target per local warehouse. This model may be denoted as a METRIC model (cf. Sherbrooke [13, 14]; notice, however, that in the standard METRIC model, one has one service target for all local warehouses together instead of a target per local warehouse). For the performance evaluation under given basestock levels, we present and compare an exact evaluation procedure, an approximate method based on two-moment fits, and the so-called METRIC approach (which is based on single-moment fits). And, we consider a heuristic method to obtain near-optimal basestock levels. Next, in Chap. 7, we generalize these methods for a multi-echelon, multi-indenture model. The models of this third block may be directly applicable in situations without a structural use of emergency and lateral shipments. For situations with a structural use of emergency and lateral shipments, they may be useful as a building block to find the right tradeoff between stocks at different echelons and/or indenture levels.

Chapter 8 forms the fourth block. In all earlier models, repair and procurement leadtimes are exogenous, and one finds tradeoffs between: (i) stocks of cheap and expensive items; (ii) stocks at different echelon levels; (iii) stocks at different indenture levels. One can also consider which parts should get priority in repair shops or which parts should get shorter procurement leadtimes. This is another dimension in which the right balance should be found to get the lowest total system costs. In Chap. 8, we look at this issue in a simplified setting with multiple repairable parts, one repair shop and one stockpoint. We show that total costs may be significantly reduced by setting the right repair priorities if one has large price differences among the repairables.

References

1. AberdeenGroup: The service parts management solution selection report, SPM strategy and technology selection handbook. AberdeenGroup, Boston (2005)
2. Basten, R.J.I., Van Houtum, G.J.: System-oriented inventory models for spare parts. Surv. Oper. Res. Manag. Sci. **19**, 34–55 (2014)

3. Cohen, M.A., Agrawal, N., Agrawal, V.: Winning in the aftermarket. Harv. Bus. Rev. **84** (May issue), 129–138 (2006)
4. Deloitte (Koudal, P.): The service revolution in global manufacturing industries. Deloitte Research (2006)
5. Dinesh Kumar, U., Crocker, J., Knezevic, J., El-Haram, M.: Reliability, Maintenance, and Logistics Support; A Life Cycle Approach. Kluwer Academic/Indian Institute of Management, Calcutta (2000)
6. Flint, P.: Too much of a good thing, better inventory management could save the industry millions while improving reliability. Air Transp. World **32**, 103–107 (1995)
7. Lam, M.: An introduction to airline maintenance. In: Handbook of Airline Economics. Aviation Week Group, New York (1995)
8. Muckstadt, J.A.: Analysis and Algorithms for Service Parts Supply Chains. Springer, New York (2005)
9. Oliva, R., Kallenberg, R.: Managing the transition from products to services. Int. J. Serv. Ind. Manag. **14**, 160–172 (2003)
10. Öner, K.B., Franssen, R., Kiesmüller, G.P., Van Houtum, G.J.: Life cycle costs measurement of complex systems manufactured by an engineer-to-order company. In: Qui, R.G., Russell, D.W., Sullivan, W.G., Ahmad, M. (eds.) The 17th International Conference on Flexible Automation and Intelligent Manufacturing, FAIM, Philadelphia, pp. 589–596 (2007)
11. Rustenburg, W.D.: A system approach to budget-constrained spare parts management. Ph.D. thesis, Eindhoven University of Technology (2000). http://w3.tue.nl/en/services/library/digilib/publications_from_tue/dissertations/
12. Rustenburg, W.D., Van Houtum, G.J., Zijm, W.H.M.: Spare parts management at complex technology-based organizations: an agenda for research. Int. J. Prod. Econ. **71**, 177–193 (2001)
13. Sherbrooke, C.C.: METRIC: a multi-echelon technique for recoverable item control. Oper. Res. **16**, 122–141 (1968)
14. Sherbrooke, C.C.: Optimal Inventory Modeling of Systems: Multi-echelon Techniques. Kluwer Academic, Boston/Dordrecht/London (2004)
15. Thonemann, U.W., Brown, A.O., Hausman, W.H.: Easy quantification of improved spare parts inventory policies. Manag. Sci. **48**, 1213–1225 (2002)
16. Wong, H., Van Houtum, G.J., Cattrysse, D., Van Oudheusden, D.: Multi-item spare parts systems with lateral transshipment and waiting time constraints. Eur. J. Oper. Res. **171**, 1071–1093 (2006)

Chapter 2
Basic Multi-Item, Single-Location Inventory Model

2.1 Introduction

In this chapter, we consider a basic multi-item, single-location inventory model for spare parts. This model is appropriate to demonstrate various optimization methods, to describe multiple system-oriented service measures, and to show the effects of alternative or additional model assumptions. The basic model is appropriate for planning in stationary situations. It may be used for both initial supply and inventory planning during the exploitation phase. For the initial supply problem, one has no starting stock, which leads to a cleaner problem formulation than for planning during the exploitation phase. Therefore, we will describe the initial supply problem first, and later we explain how this model can be used for the inventory planning during the exploitation phase.

The single-location model will fit in case one has only one location where spare parts are stocked. But, the model may also be used as a building block in planning concepts for spare parts networks (for both initial supply and the planning during the exploitation phase). This model may fit for:

- A central warehouse in a two-echelon network;
- A local warehouse;
- The total or aggregate stock in a two-echelon network with one central depot and multiple local warehouses when all these stockpoints are at closed distance of each other and emergency/lateral shipments are possible between each pair of stockpoints (in this case, one can operate as if the network as a whole forms one large virtual stockpoint).

This chapter is organized as follows. We will start with the description of the basic model under the assumption of backordered demands, i.e., without the assumption of emergency shipments. This means that a demand is backordered when it can not be satisfied from stock. In line with this assumption we will start with a service level constraint in terms of the aggregate mean number of backordered

G.J. van Houtum, B. Kranenburg, *Spare Parts Inventory Control under System Availability Constraints*, International Series in Operations Research & Management Science 227, DOI 10.1007/978-1-4899-7609-3_2

demands. For the optimization within this basic model, we will use a greedy heuristic. The description of the basic model, its evaluation, and its optimization are given in Sects. 2.2–2.4, respectively. After that, in Sect. 2.5, we discuss two alternative optimization techniques: Lagrangian relaxation and Dantzig-Wolfe decomposition. We will see that both approaches are kind of equivalent. Subsequently, in Sect. 2.6, a comparison is made with the so-called item approach, which is a straightforward inventory optimization approach without a direct connection with system availability. Next, in Sect. 2.7, we discuss alternative service measures, among which average availability. Then the application of the model for the inventory planning during the exploitation phase is discussed in Sect. 2.8. Subsequently, in Sect. 2.9, we discuss the model with emergency shipments, and we describe the changes in the analysis and optimization approaches under this alternative assumption. It will appear that these changes are limited. In Sect. 2.10, we discuss a number of practically relevant extensions for both the case without and with the use of emergency shipments. Finally, we make concluding remarks in Sect. 2.11.

Most of the material in this chapter stems from [20].

2.2 Basic Model

Consider a *single warehouse* where several spare parts are kept on stock to serve an installed base of machines of the same type. The machines consist of multiple components, which may be classified as *critical* and *non-critical* components. When a critical component of a machine fails, the whole machine goes down, while a machine can continue its functioning (i.e., to a sufficiently large extent) upon the failure of a non-critical component. We limit ourselves to the inventories of the spare parts for the critical components. When a critical component fails in a given machine, then the failed part is replaced by a spare part from the warehouse if it is available or as soon as a spare part becomes available; i.e., we have *repair by replacement*. The failed part is returned to the warehouse and is immediately sent into repair. We assume that all critical components are repairable.

We refer to the critical components as Stock-Keeping Units (SKU's). The set of SKU's is denoted by I, and the number of SKU's is denoted by $|I|$ ($\in \mathbb{N} := \{1,2,\dots\}$). For notational convenience, the SKU's are assumed to be numbered $i = 1,2,\dots,|I|$. We assume an infinite time horizon $[0,\infty)$. For each SKU $i \in I$, demand occurs according to a Poisson process with a constant rate m_i (≥ 0). The rate m_i denotes the demand rate for all machines together. The total demand rate for all SKU's together is denoted by $M = \sum_{i \in I} m_i$ and we assume that $M > 0$. A demand is fulfilled immediately if possible, and otherwise it is backordered and fulfilled as soon as possible. Each demand is accompanied by the return of a failed part, and the failed part is immediately sent into repair. The time that a failed part is in repair is called the repair leadtime, which consists of waiting time and repair time. Repair leadtimes of parts of different SKU's are assumed to be independent and repair leadtimes of parts of the same SKU are assumed to be independent and identically distributed (i.i.d.). The mean repair leadtimes for SKU i are denoted by t_i (> 0). Because each failed part is immediately sent into repair, the inventory position of SKU i, defined as the physical stock minus backordered demand plus parts in repair, is constant. This constant amount is denoted by S_i ($\in \mathbb{N}_0 := \mathbb{N} \cup \{0\}$).

Instead of saying that each failed part is immediately sent into repair, we may also say that for each SKU the stock is controlled by a continuous-review *basestock policy*, with *basestock level* S_i for SKU i. Basestock level S_i represents the initial stock of SKU i and is a decision variable.

The price of a part of SKU i is c_i^{a} (> 0). We look at the initial supply problem at time instant $t = 0$, i.e., at the investment in spare parts at the beginning of the time horizon. The objective is to minimize the total investment subject to a constraint on the aggregate mean number of backorders. The investment in, or budget spent to, spare parts of SKU i is given by $C_i(S_i) = c_i^{\mathrm{a}} S_i$ and the total investment is given by:

$$C(\mathbf{S}) = \sum_{i \in I} C_i(S_i) = \sum_{i \in I} c_i^{\mathrm{a}} S_i,$$

where $\mathbf{S} = (S_1, \ldots, S_{|I|})$ denotes a vector consisting of all basestock levels. The *mean number of backorders* of SKU i, in steady state (i.e., at an arbitrary point in time in the long run), is denoted by $EBO_i(S_i)$. The *aggregate mean number of backorders*, in steady state, is:

$$EBO(\mathbf{S}) = \sum_{i \in I} EBO_i(S_i). \tag{2.1}$$

The target level for $EBO(\mathbf{S})$ is given by EBO^{obj} and the solution space is:

$$\mathscr{S} = \{\mathbf{S} = (S_1, \ldots, S_{|I|}) \mid S_i \in \mathbb{N}_0, \forall i \in I\}.$$

Hence, in mathematical terms, our optimization problem is as follows:

$$\begin{aligned} \text{(P)} \qquad \min \qquad & C(\mathbf{S}) \\ \text{subject to } & EBO(\mathbf{S}) \le EBO^{\mathrm{obj}}, \\ & \mathbf{S} \in \mathscr{S}. \end{aligned}$$

The optimal cost of Problem (P) is denoted by C_{P}. Problem (P) has a linear objective function, a nonlinear constraint, and integral decision variables. It thus is a *nonlinear integer programming problem*.

The mean backorder position $EBO_i(S_i)$ denotes the number of parts of SKU i that is missing in all machines of the installed base together. A part is said to be missing in case a failed part has not been replaced yet by a ready-for-use part because there was no ready-for-use spare part available. Similarly, $EBO(\mathbf{S})$ denotes the total number of missing parts in all machines together, and thus is a measures for the inconvenience due to insufficient stock of ready-for-use spare parts. The constraint on the aggregate mean number of backorders is closely related to an *availability constraint*, where the availability $A(\mathbf{S})$ denotes the fraction of machines that is not down due to a missing part, or equivalently, the fraction of time that any given machine is not down due to a missing part. See Sect. 2.7.2 for a detailed description of this relation. In short, if it hardly occurs that any machine has two or more parts missing, then

$$A(\mathbf{S}) \approx 1 - \frac{1}{Z} EBO(\mathbf{S}),$$

where Z denotes the total number of machines, and thus setting a maximum level EBO^{obj} for the aggregate mean number of backorders is equivalent to setting a minimum level $A^{\mathrm{obj}} = 1 - \frac{1}{Z} EBO^{\mathrm{obj}}$ for the availability.

In the description above, we used the terminology that is common for repairable spare parts. Nevertheless, the model is easily generalized to situations where all SKU's, or a subset of SKU's, is consumable, or where *condemnation* for repairable SKU's occurs; see Sect. 2.10.1.

2.2.1 Overview of Assumptions

We summarize and discuss the main assumptions made above:

1. *Demands for the different SKU's occur according to independent Poisson processes.*
   The assumption of independent Poisson processes is justified when a failure of a component does not lead to additional failures of other components in the same machine. In general this is true. The assumption of Poisson processes is justified either when lifetimes of components are exponential or when lifetimes are generally distributed and the number of machines that is served by the warehouse is sufficiently large.
2. *For each SKU, the demand rate is constant.*
   The single warehouse serves multiple machines. When one or more machines fail and failed parts cannot be provided immediately, then some machines may be down for a while and the demand rates for a given SKU decreases accordingly. However, when the fraction of machines that is down is always sufficiently small, either because downtimes are short in general or because downtimes occur only rarely, then the decrease in demand rate is small, and thus it is reasonable to assume a constant demand rate.
3. *Repair leadtimes for different SKU's are independent and repair leadtimes for parts of the same SKU are independent and identically distributed.*
   For repairable SKU's, this assumption is justified if *planned repair leadtimes* have been agreed with repair companies (or departments). It then is the responsibility of the repair company to meet the planned leadtimes. In practice, planned leadtimes often occur either because repair is executed by an external company or in order to decompose the inventory control from the control of the repair facilities.

4. *A one-for-one replenishment strategy is applied for all SKU's.*
   This is justified as long as there are no fixed ordering costs or fixed ordering costs are small relative to the prices of the SKU's (or, thinking of the EOQ rule, relative to price multiplied by demand rate). If fixed ordering costs are relevant, then fixed order quantities may be appropriate to assume and one may follow an (s,Q) instead of a basestock policy for each SKU. This extension is described in Sect. 2.10.3.

Example 2.1. We now describe an illustrative example that is used throughout this chapter. A manufacturer of capital goods keeps spare parts on stock in a single warehouse to support a reasonably large number of installed machines. All spare parts demands are fulfilled from this warehouse. We consider three different SKU's ($|I| = 3$). The average number of failures per year is 15 for SKU 1 ($m_1 = 15$), 5 for SKU 2 ($m_2 = 5$) and 1 for SKU 3 ($m_3 = 1$). The average repair leadtimes are equal to 2 months (or $\frac{1}{6}$ year) for all three SKU's ($t_1 = t_2 = t_3 = \frac{1}{6}$). The price of SKU 1 is 1,000 Euros ($c_1^{\rm a} = 1{,}000$), the price of SKU 2 is 3,000 Euros ($c_2^{\rm a} = 3{,}000$) and the price of SKU 3 is 20,000 Euros ($c_3^{\rm a} = 20{,}000$). It is specified that the aggregate mean number of backorders may not exceed 0.1 ($EBO^{\rm obj} = 0.1$).

2.3 Evaluation

In this section, we evaluate the steady-state behavior and the aggregate mean number of backorders $EBO(\mathbf{S})$ for a given basestock policy $\mathbf{S}$. Because parts of different SKU's have no interaction, the steady-state behavior can be evaluated per SKU. This leads to a closed-form expression for $EBO_i(S_i)$. $EBO(\mathbf{S})$ itself then follows from (2.1).

Consider an arbitrary SKU i, and assume that the basestock level S_i is given. The repair and fulfilment process of this SKU is depicted by the Petri net in Fig. 2.1. On the left-hand side in this figure, demands for ready-for-use parts, accompanied with failed parts, arrive with rate m_i. The failed parts follow the upper stream in the figure. That is, they first go into repair which takes on average t_i time units. Then they arrive in a queue with ready-for-use parts. Actually this queue represents the physical stock, also called stock on hand. The demands for ready-for-use parts follow the lower stream. That is, these requests are sent to the warehouse, where they are fulfilled immediately if there is enough stock on hand and after some delay otherwise. Delayed requests are fulfilled according to a First-Come, First-Served (FCFS) discipline. When both a request and a ready-for-use part are available, they merge (i.e., the transition on the righthand side in the figure 'fires') and leave the system.

It always holds that at least one of the two queues on the righthand side in the figure is empty. If the stock on hand is positive then there will be no requests waiting for a ready-for-use part. If the number of requests in the queue (= number of backorders) is positive, then there cannot be any part in the queue with on hand stock. The number of backorders is identical to the length of the queue with requests.

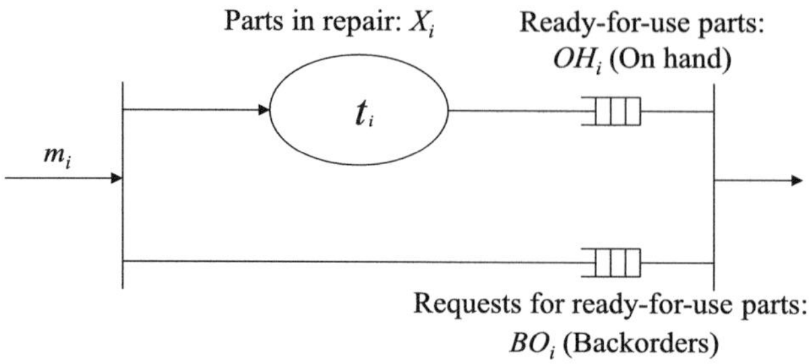

Fig. 2.1 Petri net of the repair and demand fulfilment process of SKU i

The state of the whole system at time instant t may be described by the tuples $(X_i(t), OH_i(t), BO_i(t))$, where $X_i(t)$ denotes the number of parts in repair at time t, $OH_i(t)$ denotes the stock on hand of ready-for-use parts at time t, and $BO_i(t)$ denotes the number of backordered demands at time t. The amount $X_i(t)$ represents the number of parts in the *repair pipeline* and therefore is also called the *(repair) pipeline stock.* Notice that $(X_i(t), OH_i(t), BO_i(t))$ constitutes a partial description because repair leadtimes are generally distributed, and thus for a full description one also has to denote how long parts are in repair already.

The possible values for the tuples $(X_i(t), OH_i(t), BO_i(t))$ are given by:

$$(0, S_i, 0), (1, S_i - 1, 0), \ldots, (S_i - 1, 1, 0), (S_i, 0, 0), (S_i + 1, 0, 1), (S_i + 2, 0, 2), \ldots.$$

The first S_i states in this sequence are with positive stock on hand, the state $(S_i, 0, 0)$ is the unique state where both the stock on hand and the number of backordered demands is zero, and after that the states with a positive number of backordered demands are obtained. A transition is made from one state to the next state in this sequence when a demand occurs, while a completion of a repair leads to a transition from one state to a previous state in this sequence. From the sequence with all possible states, we observe that the values of $OH_i(t)$ and $BO_i(t)$ follow directly from the value of $X_i(t)$. It holds that

$$OH_i(t) = (S_i - X_i(t))^+, \tag{2.2}$$
$$BO_i(t) = (X_i(t) - S_i)^+, \tag{2.3}$$

where $x^+ = \max\{0, x\}$ for any $x \in \mathbb{R}$. These equations imply that

$$OH_i(t) - BO_i(t) = S_i - X_i(t),$$

or, equivalently, that

$$X_i(t) + OH_i(t) - BO_i(t) = S_i.$$

This latter equation is known as the *stock balance equation* (cf. Sherbrooke [18]) and shows that the number of parts in the upper stream of the Petri net in Fig. 2.1 is always S_i more than the number of requests in the lower stream.

Let X_i, OH_i, and BO_i be the steady-state variables corresponding to $X_i(t)$, $OH_i(t)$, and $BO_i(t)$, respectively; i.e., they are random variables denoting the number of parts in repair, the number of ready-for-use parts, and the number of backordered demands in steady state. By (2.2) and (2.3),

$$OH_i = (S_i - X_i)^+, \tag{2.4}$$
$$BO_i = (X_i - S_i)^+. \tag{2.5}$$

In our model failed parts enter the repair pipeline according to a Poisson process and each failed part stays on average a time t_i in the repair pipeline. The repair pipeline may be seen as a queueing system with infinitely many servers and service times t_i. In other words, the repair pipeline is an $M|G|\infty$ queueing system and thus we may apply Palm's theorem (cf. Palm [13]):

Palm's theorem: *If jobs arrive according to a Poisson process with rate λ at a service system and if the times that the jobs remain in the service system are independent and identically distributed according to a given general distribution with mean EW, then the steady-state distribution for the total number of jobs in the service system is Poisson with mean λEW.*

Application of this theorem to the repair pipeline leads to part (i) of the following lemma; the parts (ii) and (iii) of this lemma follow from part (i) and Eqs. (2.4) and (2.5).

Lemma 2.1. *Let $i \in I$.*

(i) The pipeline stock X_i is Poisson distributed with mean $m_i t_i$, i.e.,

$$P\{X_i = x\} = \frac{(m_i t_i)^x}{x!} e^{-m_i t_i}, \qquad x \in \mathbb{N}_0;$$

(ii) The distribution of the stock on hand OH_i is given by

$$P\{OH_i = x\} = \begin{cases} \sum_{y=S_i}^{\infty} P\{X_i = y\} & \text{if } x = 0; \\ P\{X_i = S_i - x\} & \text{if } x \in \mathbb{N},\, x \le S_i; \end{cases}$$

(iii) The distribution of the number of backordered demands BO_i is given by

$$P\{BO_i = x\} = \begin{cases} \sum_{y=0}^{S_i} P\{X_i = y\} & \text{if } x = 0; \\ P\{X_i = x + S_i\} & \text{if } x \in \mathbb{N}. \end{cases}$$

Lemma 2.1 contains the main results for the evaluation of a given policy. From this lemma, we easily obtain relevant service measures, among which the mean backorder positions $EBO_i(S_i)$:

$$\begin{aligned} EBO_i(S_i) = E\{BO_i(S_i)\} &= \sum_{x=S_i+1}^{\infty} (x - S_i) P\{X_i = x\} \\ &= m_i t_i - S_i + \sum_{x=0}^{S_i} (S_i - x) P\{X_i = x\}, \qquad S_i \in \mathbb{N}_0. \end{aligned} \tag{2.6}$$

Notice that the latter expression for $EBO_i(S_i)$ is most appropriate for computational purposes as it avoids complications because of sums with infinitely many terms.

2.4 Optimization

Instead of solving Problem (P) directly, we consider a closely related Problem (Q) with two objectives, minimization of the investment $C(\mathbf{S})$ and minimization of the aggregate mean number of backorders $EBO(\mathbf{S})$:

$$
\begin{aligned}
(\mathrm{Q}) \qquad & \min \qquad C(\mathbf{S}) \\
& \min \qquad EBO(\mathbf{S}) \\
& \text{subject to } \mathbf{S} \in \mathscr{S}.
\end{aligned}
$$

This problem is a *multi-objective programming problem*. For this problem, we will derive so-called *efficient solutions*. A solution $\mathbf{S} \in \mathscr{S}$ is efficient for Problem (Q) if and only if there is no other solution $\mathbf{S}' \in \mathscr{S}$ with $C(\mathbf{S}') \leq C(\mathbf{S})$ and $EBO(\mathbf{S}') \leq EBO(\mathbf{S})$, and strict inequality for at least one of these inequalities. Alternatively stated, a solution $\mathbf{S} \in \mathscr{S}$ is efficient for Problem (Q) if and only if $C(\mathbf{S}') > C(\mathbf{S})$, or $EBO(\mathbf{S}') > EBO(\mathbf{S})$, or $(C(\mathbf{S}'), EBO(\mathbf{S}')) = (C(\mathbf{S}), EBO(\mathbf{S}))$ for all $\mathbf{S}' \in \mathscr{S}$. Let $\mathscr{E}^*$ denote the set of all efficient solutions for Problem (Q). Then the points $(C(\mathbf{S}), EBO(\mathbf{S}))$, $\mathbf{S} \in \mathscr{E}^*$, constitute an *efficient frontier* for the total inventory investment vs. aggregate mean number of backorders. From this efficient frontier, an appropriate solution for Problem (P) may be picked.

Example 2.1 (continued). For our illustrative example, we compute the mean number of backorders and inventory investment for all plausible solutions with an investment of at most 85,000 Euros. These solutions are plotted in an $C(S)$ vs. $EBO(S)$ figure; see Fig. 2.2. We are interested in the efficient solutions for problem (Q), which are denoted by squares. From this figure, we easily obtain an optimal solution of Problem (P) with $EBO^{\mathrm{obj}} = 0.1$. That solution is the first efficient solution in Fig. 2.2 with $EBO(\mathbf{S}) \leq 0.1$. This leads to the solution $\mathbf{S} = (6,2,1)$, for which $EBO(\mathbf{S}) = 0.098$ and $C(\mathbf{S}) =$ 32,000 Euros.

Problem (Q) has the following structure:

$$
\begin{aligned}
C(\mathbf{S}) &= \sum_{i \in I} C_i(S_i), \\
EBO(\mathbf{S}) &= \sum_{i \in I} EBO_i(S_i), \\
\mathscr{S} &= \mathscr{S}_1 \times \mathscr{S}_2 \times \ldots \times \mathscr{S}_{|I|},
\end{aligned}
$$

where $\mathscr{S}_i = \mathbb{N}_0$ represents the solution space for S_i for all $i \in I$, i.e., the objective functions are separable and the solutions space is a Cartesian Product, and thus Problem (Q) as a whole is *separable* (cf. Fox [8]). In addition, the functions $C_i(S_i) = c_i^{\mathrm{a}} S_i$, $i \in I$, are linear, and, as we shall derive in Sect. 2.4.1, the functions $EBO_i(S_i)$, $i \in I$, are decreasing and convex. Therefore, a greedy procedure may be applied to generate efficient solutions; see Sect. 2.4.2.

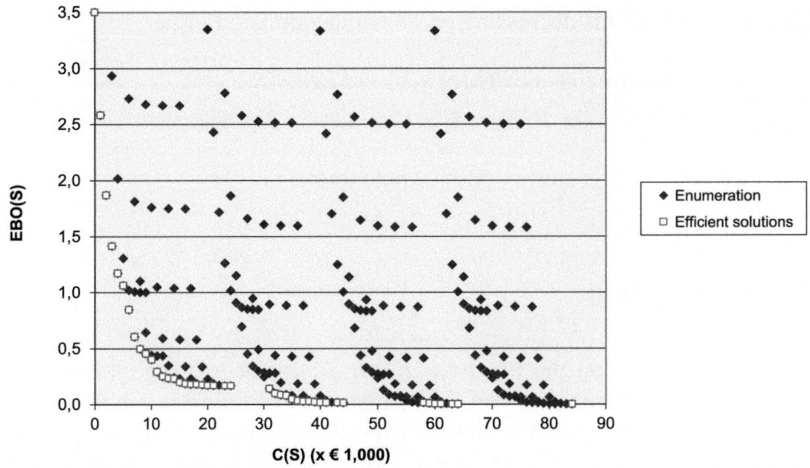

Fig. 2.2 Efficient solutions for Example 2.1

2.4.1 Convexity of the Mean Backorder Positions

Definition 2.1. Let $f(x)$ be a function on $\mathbb{Z}$, and $x_0 \in \mathbb{Z}$.

(i) $f(x)$ is decreasing for $x \geq x_0$ if

$$\Delta f(x) = f(x+1) - f(x) \leq 0, \qquad x \geq x_0;$$

(ii) $f(x)$ is convex for $x \geq x_0$ if

$$\Delta^2 f(x) = \Delta f(x+1) - \Delta f(x) \geq 0, \qquad x \geq x_0.$$

Notice that $\Delta f(x+1) - \Delta f(x) = f(x+2) - 2f(x+1) + f(x)$, $x \in \mathbb{Z}$. The definitions for strictly decreasing and strictly convex are obtained by replacing the inequality signs by strict inequality signs. The definitions for (strictly) increasing and (strictly) concave are obtained by turning the (strict) inequality signs around.

The mean number of backorders $EBO_i(S_i)$ for SKU $i \in I$ is a function on $\mathbb{N}_0$. Lemma 2.2 says that $EBO_i(S_i)$ is decreasing and convex on its whole domain.

Lemma 2.2. *For each SKU $i \in I$, $EBO_i(S_i)$ is decreasing and convex for $S_i \in \mathbb{N}_0$.*

Proof. Let $i \in I$. By (2.6),

$$\begin{aligned} \Delta EBO_i(S_i) &= EBO_i(S_i+1) - EBO_i(S_i) \\ &= -\sum_{x=S_i+1}^{\infty} P\{X_i = x\} \leq 0, \qquad S_i \in \mathbb{N}_0, \end{aligned} \tag{2.7}$$

which shows that $EBO_i(S_i)$ is decreasing on its whole domain. Further,

$$\begin{aligned} \Delta^2 EBO_i(S_i) &= \Delta EBO_i(S_i+1) - \Delta EBO_i(S_i) \\ &= P\{X_i = S_i+1\} \geq 0, \qquad S_i \in \mathbb{N}_0, \end{aligned}$$

which shows that $EBO_i(S_i)$ is convex on its whole domain. □

2.4.2 Greedy Algorithm

Problem (Q) is separable and the functions $EBO_i(S_i)$ are decreasing and convex on their whole domains. Hence we can prove that a set of efficient solutions can be generated by a greedy algorithm. A first efficient solution $\mathbf{S} = (S_1, \ldots, S_{|I|})$ is obtained by setting $S_i = 0$ for each SKU $i \in I$. This solution is efficient because it has the lowest possible investment $C(\mathbf{S}) = 0$. Next, for each SKU i, we compute the decrease in $EBO(\mathbf{S})$ relative to the increase in $C(\mathbf{S})$ when S_i would be increased by one unit. The increase in $C(\mathbf{S})$ equals c_i^{a}, while the change in $EBO(\mathbf{S})$ equals (use (2.7))

$$\Delta_i EBO(\mathbf{S}) = \Delta EBO_i(S_i) = -\sum_{x=S_i+1}^{\infty} P\{X_i = x\} = -\left(1 - \sum_{x=0}^{S_i} P\{X_i = x\}\right).$$

The decrease in $EBO(\mathbf{S})$, which is equal to $-\Delta_i EBO(\mathbf{S})$, divided by the increase in $C(\mathbf{S})$ is denoted by Γ_i. The SKU with the highest value for Γ_i is selected (also referred to as "biggest bang for the buck"), and the corresponding basestock level is increased by one unit (ties may be broken with equal probabilities). The new solution $\mathbf{S}$ is also efficient and is added to a set of efficient solutions. The generation of efficient solutions is continued until a given aggregate mean number of backorders or inventory investment has been reached, or until some other stop criterium is met. The formal procedure is described in Algorithm 2.1, where $\mathbf{e}_k$ is an $|I|$-dimensional unit row-vector.

Algorithm 2.1 *(Greedy algorithm)*

Step 1 Set $S_i := 0$ for all $i \in I$, and $\mathbf{S} = (0, 0, \ldots, 0)$;
  $\mathscr{E} := \{\mathbf{S}\}$;
  $C(\mathbf{S}) := 0$ and $EBO(\mathbf{S}) := \sum_{i \in I} m_i t_i$.
Step 2 $\Gamma_i := (1 - \sum_{x=0}^{S_i} P\{X_i = x\})/c_i^{\mathrm{a}}$ for all $i \in I$;
  $k := \arg\max\{\Gamma_i : i \in I\}$;
  $\mathbf{S} := \mathbf{S} + \mathbf{e}_k$;
  $\mathscr{E} := \mathscr{E} \cup \{\mathbf{S}\}$.
Step 3 Compute $C(\mathbf{S})$ and $EBO(\mathbf{S})$;
  If 'stop criterium', then stop, else goto Step 2.

In the following lemma, it is formally stated that Algorithm 2.1 generates a set of efficient solutions for Problem (Q). The proof of this lemma follows directly from Theorem 2 in Fox [8].

Lemma 2.3. *At termination of Algorithm 2.1, the set $\mathscr{E}$ consists of efficient solutions for Problem (Q), i.e., $\mathscr{E} \subset \mathscr{E}^*$.*

For the computation of the pipeline stock probabilities $P\{X_i = x\}$ in Algorithm 2.1, we advice to use the following recursion for the sake of efficiency and to avoid numerical problems:

$$\begin{aligned} P\{X_i = 0\} &= e^{-m_i t_i}, \\ P\{X_i = x+1\} &= \frac{m_i t_i}{x+1} P\{X_i = x\} \text{ for } x \in \mathbb{N}_0. \end{aligned} \tag{2.8}$$

Example 2.1 (continued). Application of the greedy algorithm to our example with stop criterium '$EBO(\mathbf{S}) \leq 0.1$' leads to the efficient solutions displayed in Table 2.1. After 11 iterations we obtain the first efficient solution that satisfies $EBO(\mathbf{S}) \leq 0.1$. This solution is $\mathbf{S} = (7,3,1)$, for which $EBO(\mathbf{S}) = 0.031$ and $C(\mathbf{S}) = 36{,}000$ Euros. This solution is optimal for Problem (P) with $EBO^{\mathrm{obj}} = 0.031$. Further, this solution is feasible for Problem (P) with $EBO^{\mathrm{obj}} = 0.1$, but apparently not optimal. Earlier, we found that $\mathbf{S} = (6,2,1)$ is optimal for Problem (P) with $EBO^{\mathrm{obj}} = 0.1$. Notice that the gap in costs between $\mathbf{S} = (7,3,1)$ and the optimal solution is equal to $\frac{4{,}000}{32{,}000} = 12.5\,\%$. The efficient solution $\mathbf{S} = (6,2,1)$ is not generated by the greedy algorithm. In general, the greedy algorithm generates only a subset of all efficient solutions. This follows clearly from Fig. 2.3, where both the efficient solutions from the enumeration and the efficient solutions from the greedy algorithm are displayed.

In general, the greedy algorithm generates an ordered set $\mathscr{E} = \{\mathbf{S}^0, \mathbf{S}^1, \mathbf{S}^2, \ldots\}$ of efficient solutions for Problem (Q), where $EBO(\mathbf{S}^0) > EBO(\mathbf{S}^1) > EBO(\mathbf{S}^2) > \ldots$ and $0 = C(\mathbf{S}^0) < C(\mathbf{S}^1) < C(\mathbf{S}^2) < \ldots$. The set $\mathscr{E}$ is a subset of the set $\mathscr{E}^*$ with all efficient solutions. For Problem (P) with a given target EBO^{obj}, one can easily obtain a feasible solution from the subset $\mathscr{E}$ generated by the greedy algorithm. One just takes the first solution $\mathbf{S}^l \in \mathscr{E}$ with $EBO(\mathbf{S}^l) \leq EBO^{\mathrm{obj}}$. This solution is optimal if and only if there is no solution $\mathbf{S} \in \mathscr{E}^*$ with $EBO(\mathbf{S}^l) < EBO(\mathbf{S}) \leq EBO^{\mathrm{obj}}$. In general, the solution $\mathbf{S}^l$ will be close to optimal if $EBO(\mathbf{S}^l)$ is close to EBO^{obj}. In the above example, we were a bit unlucky, because in the 11-th iteration the basestock level of SKU 3 was increased, which led to a large jump for the aggregate mean number of backorders. For real-life problems, one often has many SKU's and then such large jumps become less likely. See also the following example.

Example 2.2. The data in this example are taken from a real life situation of a repair shop which has 99 SKU's in stock. The prices of the SKU's range from 135 Euros to 61,828 Euros and the average price is 2,236 Euros. The number of failures ranges from 1 to 18 per year and the average number of failures is 2.55 per year. Figure 2.4 gives an overview of the prices and failures rates (per year) of the SKU's. The repair

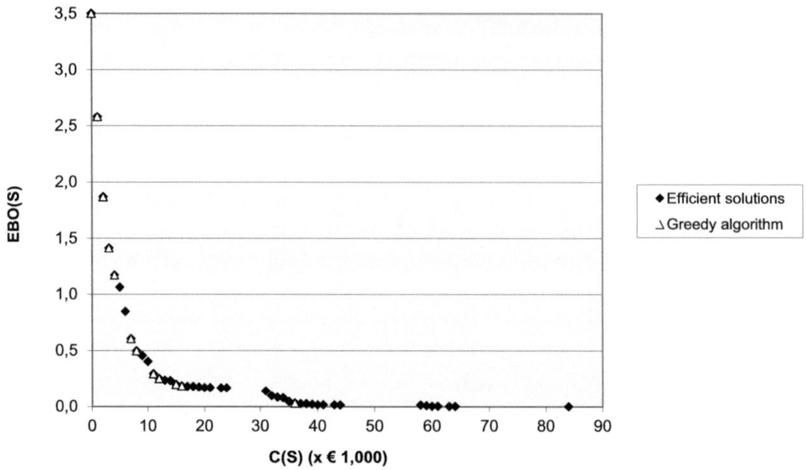

Fig. 2.3 Comparison of efficient solutions generated by the greedy algorithm and the whole set of efficient solutions

Table 2.1 Steps of the greedy algorithm for Example 2.1

Iteration	Γ_1	Γ_2	Γ_3	k	S_1	S_2	S_3	$EBO(\mathbf{S})$	$C(\mathbf{S})$ (Euros)
0	–	–	–	–	0	0	0	3.500	0
1	$9.18 \cdot 10^{-4}$	$1.88 \cdot 10^{-4}$	$7.68 \cdot 10^{-6}$	1	1	0	0	2.582	1,000
2	$7.13 \cdot 10^{-4}$	$1.88 \cdot 10^{-4}$	$7.68 \cdot 10^{-6}$	1	2	0	0	1.869	2,000
3	$4.56 \cdot 10^{-4}$	$1.88 \cdot 10^{-4}$	$7.68 \cdot 10^{-6}$	1	3	0	0	1.413	3,000
4	$2.42 \cdot 10^{-4}$	$1.88 \cdot 10^{-4}$	$7.68 \cdot 10^{-6}$	1	4	0	0	1.171	4,000
5	$1.09 \cdot 10^{-4}$	$1.88 \cdot 10^{-4}$	$7.68 \cdot 10^{-6}$	2	4	1	0	0.605	7,000
6	$1.09 \cdot 10^{-4}$	$6.77 \cdot 10^{-5}$	$7.68 \cdot 10^{-6}$	1	5	1	0	0.497	8,000
7	$4.20 \cdot 10^{-5}$	$6.77 \cdot 10^{-5}$	$7.68 \cdot 10^{-6}$	2	5	2	0	0.293	11,000
8	$4.20 \cdot 10^{-5}$	$1.74 \cdot 10^{-5}$	$7.68 \cdot 10^{-6}$	1	6	2	0	0.251	12,000
9	$1.42 \cdot 10^{-5}$	$1.74 \cdot 10^{-5}$	$7.68 \cdot 10^{-6}$	2	6	3	0	0.199	15,000
10	$1.42 \cdot 10^{-5}$	$3.47 \cdot 10^{-6}$	$7.68 \cdot 10^{-6}$	1	7	3	0	0.185	16,000
11	$4.25 \cdot 10^{-6}$	$3.47 \cdot 10^{-6}$	$7.68 \cdot 10^{-6}$	3	7	3	1	0.031	36,000

lead time is 4 months ($\frac{1}{3}$ year) for all SKU's. We are interested in the solution of Problem (P) with a target $EBO^{\text{obj}} = 3.3$ (which is comparable to the target in Example 2.1).

We applied the greedy algorithm to this data set, which led to a feasible solution $\mathbf{S}$ after 239 iterations, with $EBO(\mathbf{S}) = 3.02$ and $C(\mathbf{S}) = 277{,}749$ Euros. In Fig. 2.5, the solutions from the greedy algorithm are displayed. In this figure we see that the iterations of the greedy algorithm produce a very smooth curve. Towards the end we see somewhat bigger gaps in the line. The cost of the last but one solution was 266,593 Euros, and constitutes a lower bound for the optimal cost. Hence, we know that the generated heuristic solution is within $\frac{277{,}749-266{,}593}{266{,}593} = \frac{11{,}156}{266{,}593} = 4.2\,\%$ of

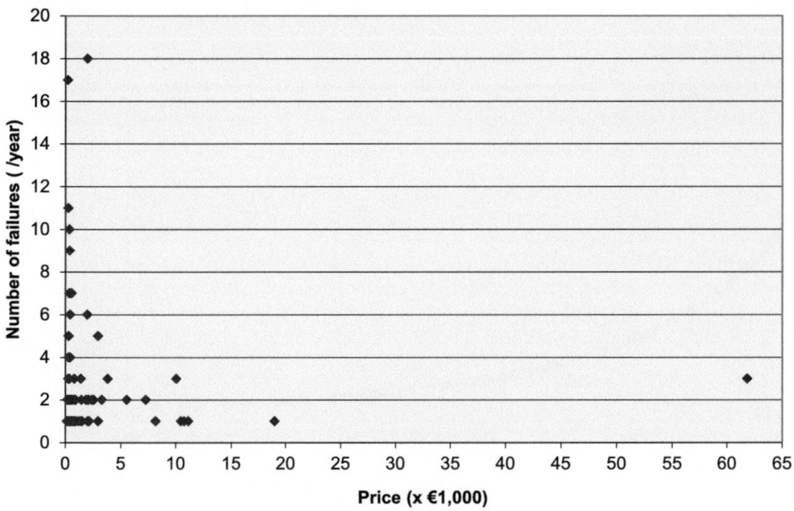

Fig. 2.4 Representation of all SKU's

the optimal solution (this is a bound, the actual gap is smaller). For applications in practice, this is sufficiently good.

We can conclude that for instances with sufficiently many SKU's, the greedy algorithms will generate good heuristic solutions for Problem (P). Besides, a greedy algorithm is efficient, it is easy to implement in practice and it is an algorithm that is easy to understand by practitioners. If one would be interested in optimal solutions, then Problem (P) may be solved by a similar exact approach as for knapsack problems. A disadvantage of such an approach is that a small change in input parameters (cost prices of the SKU's, demand rates, or the target EBO^{obj}) may lead to large changes in the optimal solution. The solution generated by the greedy algorithm, however, will be rather robust. That is another big advantage of the greedy algorithm.

2.5 Alternative Optimization Techniques

We can also approach the Problems (P) and (Q) with other techniques than the greedy algorithm of Sect. 2.4.2. In this section we will describe two alternative optimization techniques, Lagrangian relaxation is described in Sect. 2.5.1 and the Dantzig-Wolfe decomposition is described in Sect. 2.5.2.

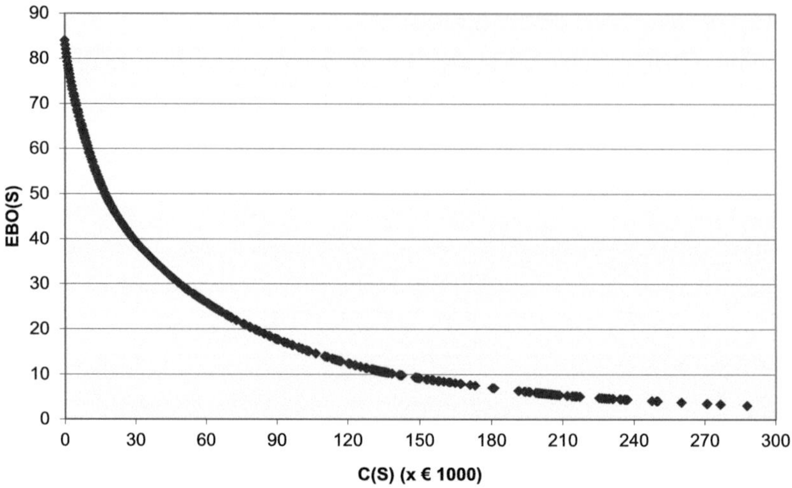

Fig. 2.5 Outcome of the greedy algorithm

2.5.1 Lagrangian Relaxation

We apply the Lagrangian relaxation technique to Problem (P); for a general description of this technique, we refer to Appendix B of Porteus [14] (other well-known references on Lagrangian relaxation are Everett [4] and Fisher [6, 7]). The *Lagrangian function* for (P) is defined as

$$L(\mathbf{S},\lambda) := \sum_{i\in I} c_i^{\mathrm{a}} S_i + \lambda \left(\sum_{i\in I} EBO_i(S_i) - EBO^{\mathrm{obj}} \right)$$

where $\lambda \geq 0$ is a *Lagrange multiplier*.

It has been noticed before that Problem (Q) is separable. This also holds for Problem (P) (see also the definition of separable problems in [14], Appendix B). We can separate Problem (P) because it is a linear combination of SKU objectives and constraints. It is known that in separable problems, the Lagrangian function is also separable. The Lagrangian function is now defined as

$$L(\mathbf{S},\lambda) = \sum_{i\in I} L_i(S_i,\lambda) - \lambda EBO^{\mathrm{obj}}, \tag{2.9}$$

where

$$L_i(S_i,\lambda) := c_i^{\mathrm{a}} S_i + \lambda EBO_i(S_i)$$

is the *decentralized Lagrangian function* for SKU i. Notice that, in Eq. (2.9), we have $|I|$ different Lagrangian functions, one for every SKU. Notice that we have only one λ because we have only one constraint in our problem.

For any given value of λ, we can easily find a base stock level that minimizes the decentralized Lagrangian function $L_i(S_i, \lambda)$, $i \in I$. Since each decentralized Lagrangian function is a convex functions, we know that it has either one unique minimum or multiple minima in subsequent points. One way to find a minimum is to start with $S_i = 0$ and increase this S_i by 1 at a time, until the value for the decentralized Lagrangian function start increasing. We can do this for all $i \in I$, and the resulting base stock vector $\mathbf{S}$ is a feasible solution for problem (P) (under any choice for EBO^{obj}). We can now vary the value of λ to find different solutions for Problem (P). Then, we calculate the corresponding values of $EBO(\mathbf{S})$ and $C(\mathbf{S})$.

Example 2.1 (continued). Applying Lagrangian relaxation to our problem, gives the following decentralized Lagrangian functions:

$$L_i(S_i, \lambda) = \begin{cases} 1{,}000 \cdot S_1 + \lambda EBO_1(S_1) & \text{if } i = 1 \\ 3{,}000 \cdot S_2 + \lambda EBO_1(S_2) & \text{if } i = 2 \\ 20{,}000 \cdot S_3 + \lambda EBO_1(S_3) & \text{if } i = 3. \end{cases}$$

If we vary λ from 0 to 300,000, we find the solutions that are displayed in Table 2.2. We can see that Lagrangian relaxation yields exactly the same solutions as our greedy algorithm (compare Tables 2.1 and 2.2). In each row of Table 2.2, a range of values for λ is given for which that specific base stock vector minimizes the Lagrangian function.

In this example, we observe a few important properties. The first property is that using the Lagrangian relaxation method gives us optimal solutions of Problem (P) for specific values of EBO^{obj}. This follows from the so-called Everett result (Everett [4]), which for our problem reads as follows:

The Everett result: *If, for a given $\lambda \geq 0$, $\mathbf{S}(\lambda)$ minimizes $L(\mathbf{S}, \lambda)$ over $\mathbf{S} \in \mathscr{S}$, then $\mathbf{S}(\lambda)$ is optimal for Problem (P) for every $EBO^{\text{obj}} \in (0, \infty)$ that satisfies*

$$EBO^{\text{obj}} \geq EBO(\mathbf{S}(\lambda)) \text{ and } \lambda\,(EBO(\mathbf{S}(\lambda))-EBO^{\text{obj}}) = 0.$$

If we take $\lambda = 0$, then each Lagrangian function $L_i(S_i, \lambda)$ is strictly increasing, and we find $\mathbf{S}(0) = (0, \ldots, 0)$ and $EBO(\mathbf{S}(0)) = \sum_{i \in I} m_i t_i$. The solution $\mathbf{S}(0) = (0, \ldots, 0)$ is optimal for Problem (P) for every $EBO^{\text{obj}} \geq \sum_{i \in I} m_i t_i$. For each $\lambda > 0$, the solution $\mathbf{S}(\lambda)$ is optimal for Problem (P) for $EBO^{\text{obj}} = EBO(\mathbf{S}(\lambda))$; i.e., then the optimality of $\mathbf{S}(\lambda)$ is guaranteed for one specific value of EBO^{obj} (but the solution might also be optimal for slightly lower values of EBO^{obj}).

The second property that we see back in the above example is that the Lagrangian relaxation method gives efficient solutions for Problem (Q). This follows directly from Theorem 1 in Fox [8].

A third property that we observe is that Lagrangian relaxation yields exactly the same solutions as the greedy algorithm. This is not a coincidence. When we study the details of the execution of the greedy algorithm and the execution of the Lagrangian relaxation method, we see the similarities. The key is that a one-to-one

Table 2.2 Solutions generated with Lagrangian relaxation

λ	S_1	S_2	S_3	$EBO_1(S_1)$	$EBO_2(S_2)$	$EBO_3(S_3)$	$EBO(\mathbf{S})$	$C(\mathbf{S})$ (Euros)
$\in [0,1089)$	0	0	0	2.500	0.833	0.167	3.500	0
$\in [1{,}089,1{,}403)$	1	0	0	1.582	0.833	0.167	2.582	1,000
$\in [1{,}403,2{,}192)$	2	0	0	0.869	0.833	0.167	1.869	2,000
$\in [2{,}192,4{,}125)$	3	0	0	0.413	0.833	0.167	1.413	3,000
$\in [4{,}125,5{,}306)$	4	0	0	0.171	0.833	0.167	1.171	4,000
$\in [5{,}306,9{,}189)$	4	1	0	0.171	0.268	0.167	0.605	7,000
$\in [9{,}189,14{,}761)$	5	1	0	0.062	0.268	0.167	0.497	8,000
$\in [14{,}761,23{,}798)$	5	2	0	0.062	0.065	0.167	0.293	11,000
$\in [23{,}798,57{,}324)$	6	2	0	0.020	0.065	0.167	0.251	12,000
$\in [57{,}324,70{,}486)$	6	3	0	0.020	0.012	0.167	0.199	15,000
$\in [70{,}486,130{,}278)$	7	3	0	0.006	0.012	0.167	0.185	16,000
$\in [130{,}278,287{,}985)$	7	3	1	0.006	0.012	0.013	0.031	36,000

relationship exists between the Γ_i values computed in the greedy algorithm and the values of λ for which $\mathbf{S}(\lambda)$ changes to the next solution in the Lagrangian relaxation method.

2.5.2 Dantzig-Wolfe Decomposition

A method that one may also apply to solve Problem (P) is Dantzig-Wolfe decomposition, as introduced by Dantzig and Wolfe [3]. In order to be able to apply Dantzig-Wolfe decomposition to Problem (P), we need to redefine it. A *Master Problem* is introduced in which all possible basestock levels per SKU are listed as columns. A constraint is added to ensure that only one basestock level is chosen per SKU. Set $K := \mathbb{N}_0$ contains all possible basestock levels for all SKU's i, $i \in I$. Let S_i^k, $i \in I$, $k \in K$, be a variable referring to a fixed policy k for SKU i. We introduce a new variable $x_i^k \in \{0,1\}$, $i \in I$, $k \in K$, that indicates whether a specific policy is selected for item i or not. For example, if $x_i^3 = 1$ then policy 3 is selected for SKU i, this is equivalent with $S_i = 2$. By relaxing the integrality constraint on x_i^k, we allow for randomized policies. The Master Problem (MP) related to Problem (P) is now defined as follows:

$$
\begin{aligned}
\text{(MP)} \qquad & \min \sum_{i\in I}\sum_{k\in K} C_i(S_i^k)x_i^k \\
& \text{subject to } \sum_{i\in I}\sum_{k\in K} EBO_i(S_i^k)x_i^k \le EBO^{\text{obj}} \\
& \sum_{k\in K} x_i^k = 1, \qquad i \in I \\
& x_i^k \ge 0, \qquad i \in I,\ k \in K.
\end{aligned}
$$

Note that the costs $C_i(S_i^k)$ and average backorder ($EBO_i(S_i^k)$) are pre-determined for all basestock levels S_i^k and that we use these numbers as input for Problem (MP). This Master Problem can have many input variables, or possible policies. The size of the problem may restrict fast calculation of the solution. Therefore, a *Restricted Master Problem* (RMP) is introduced that considers only a subset of all possible policies ($K_i \subseteq K$) per SKU i and therefore requires less computational effort:

$$\text{(RMP)} \qquad \min \sum_{i \in I} \sum_{k \in K_i} C_i(S_i^k) x_i^k$$

$$\text{subject to} \sum_{i \in I} \sum_{k \in K_i} EBO_i(S_i^k) x_i^k \leq EBO^{\text{obj}}$$

$$\sum_{k \in K_i} x_i^k = 1, \qquad i \in I$$

$$x_i^k \geq 0, \qquad i \in I,\ k \in K_i.$$

Initially, the Problem (RMP) will be solved with one policy per SKU ($|K_i| = 1$). Thereafter, we will employ a procedure called column generation to add policies, or columns, to the policy set K_i. We define an initial solution that is feasible but not optimal. Let K_i initially consist of one policy k that is defined by $S_i^k := \min\{S_i | EBO_i(S_i) \leq \frac{m_i}{M} EBO^{\text{obj}}, S_i \in \mathbb{N}_0\}$.

From the optimal solution of the (RMP) we also obtain the dual prices of all constraints. Let u be the dual price that corresponds to the EBO constraint and let v_i be the dual price that corresponds to the constraint that ensures that only one policy is chosen for SKU i. Dual price u needs to be nonpositive and v_i is unrestricted in sign.

In the column generation *subproblem* of SKU i, we search for policies that have not yet been considered and improve our solution for (RMP), and these policies are then added to K_i. To obtain these policies we need to solve a subproblem for each SKU i. Subproblem i is defined as follows:

$$(\text{SUB}(i)) \qquad \min C_i(S_i) - EBO_i(S_i)u - v_i$$

$$\text{subjext to } S_i \in \mathbb{N}_0.$$

We are only interested in a basestock level that minimizes the subproblem and that has a so-called negative reduced cost coefficient (i.e., under this basestock level, the objective function of (SUB(i)) has a negative value). We know that the solution of (SUB(i)) is straightforward because its objective function is the sum of a convex term ($-EBO_i(S_i)u$), a linear term ($C_i(S_i)$) and a constant term ($-v_i$). After solving the subproblem (SUB(i)), we add the obtained optimal policy to the set K_i if it has a negative reduced cost. We do this for all subproblems (SUB(i)), $i \in I$. We then solve the (RMP) again for the extended set of policies and use the new dual prices as input for the subproblems (SUB(i)). We continue this until all subproblems (SUB(i)) have a nonnegative reduced cost coefficient.

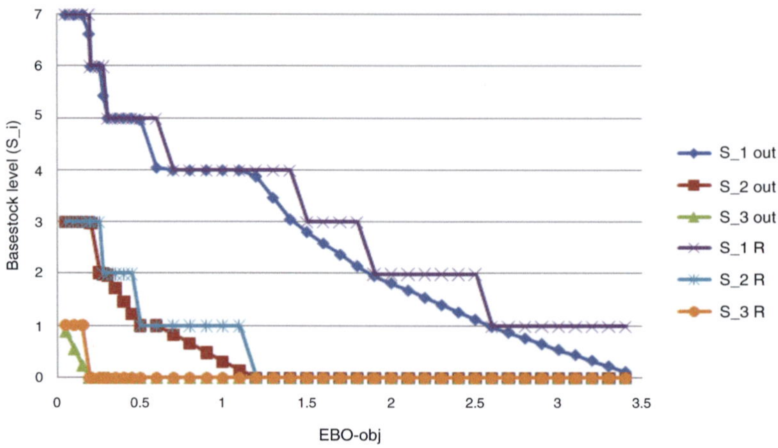

Fig. 2.6 Basestock vectors generated with Dantzig-Wolfe decomposition

One final step in the Dantzig-Wolfe method is to determine the basestock vector that corresponds to the optimal solution for Problem (P). If the solution that we obtain from the (RMP) consists of some fractional values, we need to round up these basestock levels to the nearest integer. It appears that at most one SKU has a fractional base stock value.

Example 2.1 (continued). We applied Dantzig-Wolfe decomposition to the example. For different value of EBO^{obj} we determined the optimal basestock level. As said before, rounding up was used to obtain integer values for basestock levels. The results of this analysis are graphically displayed in Fig. 2.6. For varying levels of EBO^{obj}, we have depicted the value of the basestock levels for all SKU's (indicated by "S_i out"), and the rounded values (indicated by "S_i R"). (A basestock value of 2.5 of SKU i corresponds to $x_i^2 = x_i^3 = 0.5$.) From this figure, we see that the fractional basestock levels decrease almost linearly in the target EBO between two integer values.

The results are aggregated in Table 2.3. If we compare Tables 2.3 and 2.2, we see that the solutions we obtain are equal. Apparently for this problem, the two optimization methods yield exactly the same solutions. This fact also implies that with Dantzig-Wolfe decomposition we obtain the same results as with the Greedy Algorithm. That we obtain the same results for Dantzig-Wolfe decomposition and Lagrangian relaxation is due to the strong similarities between these two methods; see also [10] (Sect. 1.4.2).

Table 2.3 Solutions generated with Dantzig-Wolfe decomposition

EBO^{obj}	S_1	S_2	S_3	$EBO_1(S_1)$	$EBO_2(S_2)$	$EBO_3(S_3)$	$EBO(\mathbf{S})$	$C(\mathbf{S})$ (Euros)
≥ 3.500	0	0	0	2.500	0.833	0.167	3.500	0
$\in [3.500, 2.582)$	1	0	0	1.582	0.833	0.167	2.582	1,000
$\in [2.582, 1.869)$	2	0	0	0.869	0.833	0.167	1.869	2,000
$\in [1.869, 1.413)$	3	0	0	0.413	0.833	0.167	1.413	3,000
$\in [1.413, 1.171)$	4	0	0	0.171	0.833	0.167	1.171	4,000
$\in [1.171, 0.605)$	4	1	0	0.171	0.268	0.167	0.605	7,000
$\in [0.605, 0.497)$	5	1	0	0.062	0.268	0.167	0.497	8,000
$\in [0.497, 0.293)$	5	2	0	0.062	0.065	0.167	0.293	11,000
$\in [0.293, 0.251)$	6	2	0	0.020	0.065	0.167	0.251	12,000
$\in [0.251, 0.199)$	6	3	0	0.020	0.012	0.167	0.199	15,000
$\in [0.199, 0.185)$	7	3	0	0.006	0.012	0.167	0.185	16,000
$\in [0.185, 0.031)$	7	3	1	0.006	0.012	0.013	0.031	36,000

2.6 Item Approach

So far, we treated all SKU's in one model and thus we followed a so-called system approach (cf. Sect. 1.3). A straightforward way to get a feasible solution for Problem (P) is to decompose the constraint for the aggregate mean number of backorders into constraints per SKU. One then gets a simple decision problem per SKU. This simplified approach is a so-called *item approach* and is obviously suboptimal. In this section, we describe the item approach and we compare its solution for Problem (P) to the solutions found by the methods of the previous sections.

The item approach consists of the following steps:

1. Set a target EBO^{obj} for the aggregate mean number of backorders;
2. Divide this target over the SKU's based on the demand rates; i.e., set a target EBO_i^{obj} for the mean number of backorders of SKU i via the following formula: $EBO_i^{\text{obj}} = (m_i/M)EBO^{\text{obj}}$;
3. For each SKU i: determine S_i such that the target EBO_i^{obj} is reached against minimal costs; i.e., determine S_i as the smallest S_i for which $EBO_i(S_i) \leq EBO_i^{\text{obj}}$;
4. Determine $EBO(\mathbf{S})$ and $C(\mathbf{S})$.
   Because all S_i are integer valued, the actual $EBO_i(S_i)$ can be significantly smaller than the corresponding target EBO_i^{obj} for a SKU i, and thus the actual $EBO_i(\mathbf{S})$ can be quite a bit smaller than the aggregate target EBO^{obj}. By varying EBO^{obj}, we obtain an $EBO(\mathbf{S})$ versus $C(\mathbf{S})$ curve, as under the approaches of Sects. 2.4 and 2.5. Via this curve, a heuristic solution is obtained for Problem (P) with a given target level EBO^{obj}.

Example 2.1 (continued). We apply the item approach to our problem for varying levels of the target EBO^{obj}, which leads to the solutions displayed in Table 2.4 and Fig. 2.7. The generated solutions are different from the solutions that we obtained before. For $EBO^{\text{obj}} = 0.1$, the item approach leads to the solution $(5, 3, 2)$ and a

Table 2.4 Inventory investment for different values of target item mean backorder

EBO^{obj}	EBO_1^{obj}	EBO_2^{obj}	EBO_3^{obj}	S_1	S_2	S_3	$EBO(\mathbf{S})$	$C(\mathbf{S})$ (Euros)
0.03	0.02	0.01	0.00	6	4	2	0.02	58,000
0.08	0.05	0.02	0.00	6	3	2	0.03	55,000
0.15	0.11	0.04	0.01	5	3	2	0.09	54,000
0.20	0.14	0.05	0.01	5	3	2	0.09	54,000
0.30	0.21	0.07	0.01	4	2	1	0.25	30,000
0.60	0.43	0.14	0.03	3	2	1	0.49	29,000
0.90	0.64	0.21	0.04	3	2	1	0.49	29,000
1.20	0.86	0.29	0.06	3	1	1	0.69	26,000
1.50	1.07	0.36	0.07	2	1	1	1.15	25,000
1.80	1.29	0.43	0.09	2	1	1	1.15	25,000
2.10	1.50	0.50	0.10	2	1	1	1.15	25,000
2.40	1.71	0.57	0.11	1	1	1	1.86	24,000
2.70	1.93	0.64	0.13	1	1	1	1.86	24,000

corresponding investment of 54,000 Euros. We can see from the graph that the item approach does not work well. In general, the item approach works well if the differences between the prices of the SKU's are small. The solutions will deviate considerably from the efficient solutions under large price differences.

2.7 Alternative Service Measures

Now consider the situation that we are not interested in the aggregate mean number of backorders but in some other service measure. Fortunately, it is possible to adjust the greedy algorithm to deal with a different service measure. In Sects. 2.7.1 and 2.7.2, we discuss the implications of using the aggregate mean waiting time and availability, respectively, as the relevant service measure. Furthermore, we deal with the sum of the backorder probabilities in Sect. 2.7.3. In Sect. 2.7.4, we show how to deal with the aggregate fill rate. Lastly, in Sect. 2.7.5, we describe the implications of using the aggregate mean number of stockouts as a service measure.

2.7.1 Aggregate Mean Waiting Time

A service measure that is quite often considered is the expected waiting time until an arbitrary spare part demand is fulfilled. This measure, which we denote as the *aggregate mean waiting time*, is obtained as follows. Let $W_i(S_i)$ be the mean waiting time for a spare part demand of SKU $i \in I$ under base stock level S_i. Then the aggregate mean waiting time $W(\mathbf{S})$ is equal to

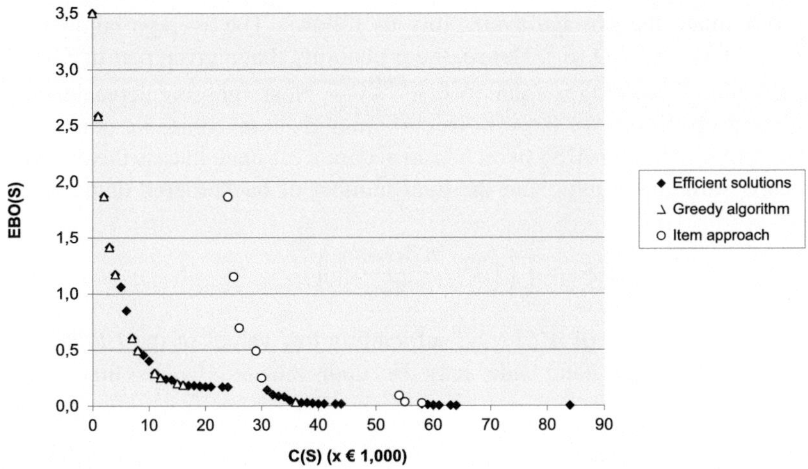

Fig. 2.7 Outcome of the item approach

$$W(\mathbf{S}) = \sum_{i \in I} P\{\text{an arbitrary demand is for SKU } i\} \times (\text{expected waiting time for SKU } i) = \sum_{i \in I} \frac{m_i}{M} W_i(S_i).$$

In Problem (P), the constraint $EBO(\mathbf{S}) \leq EBO^{\mathrm{obj}}$ is replaced by the constraint $W(\mathbf{S}) \leq W^{\mathrm{obj}}$, where W^{obj} is the target level for the aggregate mean waiting time.

By Little's law [12], it is easily seen that $W_i(S_i) = EBO_i(S_i)/m_i$, $i \in I$. As a result,

$$W(\mathbf{S}) = \sum_{i \in I} \frac{m_i}{M} \frac{EBO_i(S_i)}{m_i} = \frac{1}{M} \sum_{i \in I} EBO_i(S_i) = \frac{1}{M} EBO(\mathbf{S}),$$

and thus the constraint $W(\mathbf{S}) \leq W^{\mathrm{obj}}$ is equivalent to the constraint $EBO(\mathbf{S}) \leq EBO^{\mathrm{obj}}$ with $EBO^{\mathrm{obj}} = MW^{\mathrm{obj}}$ (notice that the relation $W(\mathbf{S}) = EBO(\mathbf{S})/M$ also is obtained by applying Little's formula to the total number of backordered demands). This means that the problem with an aggregate mean waiting time constraint can be solved in the same way as Problem (P) of the basic model.

2.7.2 Average Availability

As stated in Sect. 2.2, the constraint on the aggregate mean number of backorders is closely related to an *availability constraint*. The average availability is equal to the fraction of time that any given machine is available. Let Z be the total number of machines, and let Z_i be the number of parts of SKU i installed per machine.

We can approximate the average availability as follows. The average number of backorders of SKU i is $EBO_i(S_i)$. Hence, the probability that a given part of SKU i in a given machine is working is equal to $1-\frac{EBO_i(S_i)}{ZZ_i}$. Next, ignoring dependencies between these probabilities for the various parts in a given machine, we obtain the following approximation for $A(\mathbf{S})$ (which is an accurate estimate in case the number of machines is sufficiently large and the total number of backordered demands is low):

$$A(\mathbf{S}) \approx \prod_{i\in I}\left(1-\frac{EBO_i(S_i)}{ZZ_i}\right)^{Z_i}.$$

For sufficiently high values of $A(S)$ (i.e., sufficiently low values of the $EBO_i(S_i)$), the product on the right hand side may be approximated by its first order approximation:

$$\begin{aligned} A(\mathbf{S}) &\approx \prod_{i\in I}\left(1-\frac{EBO_i(S_i)}{ZZ_i}\right)^{Z_i} \\ &= 1-\sum_{i\in I} Z_i\frac{EBO_i(S_i)}{ZZ_i}+\sum_{i\in I}\frac{Z_i(Z_i-1)}{2}\left(\frac{EBO_i(S_i)}{ZZ_i}\right)^2 \\ &\quad +\sum_{i,j\in I,\ i\neq j} Z_iZ_j\frac{EBO_i(S_i)}{ZZ_i}\frac{EBO_j(S_j)}{ZZ_j}-\cdots \\ &\approx 1-\sum_{i\in I} Z_i\frac{EBO_i(S_i)}{ZZ_i}=1-\frac{1}{Z}\sum_{i\in I} EBO_i(S_i)=1-\frac{1}{Z}EBO(\mathbf{S}). \end{aligned}$$

Hence, for a sufficiently high target A^{obj} for $A(\mathbf{S})$, a heuristic solution for the problem with a target average availability may be obtained via the heuristic solution for Problem (P) with target $EBO^{\text{obj}} = Z(1-A^{\text{obj}})$ for the aggregate mean number of backorders.

2.7.3 Sum of Backorder Probabilities

In our basic model, we assumed that a relatively large group of machines is supported by the single stockpoint. In some applications, only one machine is supported. This situation occurs on board of a frigate, or when customers have one machine only and spare parts are kept on stock at the customer itself. The corresponding optimization problem has been studied by Rustenburg [15]. In this problem, the availability of the machine is equal to $A(\mathbf{S}) = \prod_{i\in I}(1-PBO_i(S_i))$, where $PBO_i(S_i)$ denotes the backorder probability of SKU i. This is the steady-state probability for a positive number of backorders for SKU i. This probability is equal to:

$$PBO_i(S_i) = \sum_{x=S_i+1}^{\infty} P\{X_i = x\}.$$

For sufficiently high values of the availability (i.e., sufficiently low values of the $PBO_i(S_i)$), it holds that the availability is well approximated by its first order approximation

$$\begin{aligned} A(\mathbf{S}) &= 1 - \sum_{i \in I} PBO_i(S_i) + \sum_{i,j \in I,\, i \neq j} PBO_i(S_i) PBO_j(S_j) - \ldots \\ &\approx 1 - \sum_{i \in I} PBO_i(S_i), \end{aligned}$$

and then aiming for a high availability is almost equivalent to aiming for a low sum of backorder probabilities:

$$PBO(\mathbf{S}) := \sum_{i \in I} PBO_i(S_i).$$

Then we get Problem (P) with a constraint for $PBO(\mathbf{S})$. The greedy algorithm can be slightly modified to solve this problem.

For the first order difference function of $PBO_i(S_i)$, $i \in I$, we find that

$$\Delta PBO_i(S_i) = PBO_i(S_i + 1) - PBO_i(S_i) = -P\{X_i = S_i + 1\} \leq 0, \qquad S_i \in \mathbb{N}_0;$$

i.e., $PBO_i(S_i)$ is decreasing on its whole domain. Via the second order difference function of $PBO_i(S_i)$, one can show that $PBO_i(S_i)$ is convex for $S_i \geq \max\{\lceil m_i t_i - 2\rceil, 0\}$ (see Problem 2.4). The amount $m_i t_i$ represents the average number of parts in the repair pipeline. If this average pipeline stock is smaller than or equal to 2, then $\max\{\lceil m_i t_i - 2\rceil, 0\} = 0$ and thus $PBO_i(S_i)$ is convex on its whole domain. If the average pipeline stock is larger than 2 (for slow moving SKU's this will not happen in general), then $\max\{\lceil m_i t_i - 2\rceil, 0\} > 0$ and we exclude solutions $\mathbf{S}$ with $S_i \leq \max\{\lceil m_i t_i - 2\rceil, 0\}$ from the solution space. The excluded solutions have a high value for $PBO_i(S_i)$ and thus also for $PBO(\mathbf{S})$. Hence, they are not feasible for the relevant problem instances with a low target for the sum of backorder probabilities.

In the greedy algorithm, the exclusion of solutions $\mathbf{S}$ with $S_i \leq \max\{\lceil m_i t_i - 2\rceil, 0\}$ for some $i \in I$ simply implies that we use the solution $\mathbf{S}$ with $S_i = \max\{\lceil m_i t_i - 2\rceil, 0\}$ for all $i \in I$ as the starting solution. Next, in each iteration of the greedy algorithm, the ratio of the decrease in $PBO(\mathbf{S})$ and the increase in $C(\mathbf{S})$, due to an increase of the basestock level of SKU i by 1, is measured by Γ_i, which equals:

$$\Gamma_i = \frac{P\{X_i = S_i + 1\}}{c_i^{\mathrm{a}}}.$$

In each iteration, we increase the basestock level of the SKU with the highest Γ_i.

2.7.4 Aggregate Fill Rate

The *aggregate fill rate* is defined as the probability that an arbitrary demand for the total group of SKU's is fulfilled immediately, or, equivalently, as the fraction of the total demand stream that is fulfilled from stock. Let the fill rate for SKU i, also called *item fill rate*, be denoted by $\beta_i(S_i)$, then

$$\beta(\mathbf{S}) = \sum_{i \in I} \frac{m_i}{M} \beta_i(S_i). \tag{2.10}$$

The target aggregate fill rate is given by β^{obj}.

Demand for SKU i arrives according to a Poisson process, and thus, by the PASTA (Poisson Arrivals See Time Averages) property, an arbitrary arriving demand observes the system in steady state. Hence, with probability $P\{OH_i > 0\} = P\{X_i < S_i\}$, a positive stock on hand is observed and the demand can be fulfilled immediately, and otherwise not. So,

$$\beta_i(S_i) = \sum_{x=0}^{S_i-1} P\{X_i = x\}. \tag{2.11}$$

The item fill rate $\beta_i(S_i)$ for an SKU $i \in I$ is a function with domain $\mathbb{N}_0$. Lemma 2.4 says that $\beta_i(S_i)$, and thus also $f_i(S_i) = \frac{m_i}{M}\beta_i(S_i)$, is increasing on its whole domain and concave for $S_i \geq \max\{\lceil m_i t_i - 1 \rceil, 0\}$, where $\lceil x \rceil$ denotes the rounded value to above for any $x \in \mathbb{R}$.

Lemma 2.4. *For each SKU $i \in I$, the item fill rate $\beta_i(S_i)$ is increasing on its whole domain and concave for $S_i \geq \max\{\lceil m_i t_i - 1 \rceil, 0\}$.*

Proof. Let $i \in I$. By (2.11),

$$\Delta\beta_i(S_i) = \beta_i(S_i+1) - \beta_i(S_i) = P\{X_i = S_i\} \geq 0, \qquad S_i \in \mathbb{N}_0, \tag{2.12}$$

which shows that $\beta_i(S_i)$ is increasing on its whole domain. Further,

$$\Delta^2\beta_i(S_i) = P\{X_i = S_i+1\} - P\{X_i = S_i\}, \qquad S_i \in \mathbb{N}_0. \tag{2.13}$$

By Eq. (2.8),

$$P\{X_i = S_i+1\} = \frac{m_i t_i}{S_i+1} P\{X_i = S_i\}, \qquad S_i \in \mathbb{N}_0,$$

and by substitution of this recursive relation into (2.13), we find

$$\Delta^2\beta_i(S_i) = \left(\frac{m_i t_i}{S_i+1} - 1\right) P\{X_i = S_i\}, \qquad S_i \in \mathbb{N}_0.$$

From this formula, it follows that $\Delta^2\beta_i(S_i) \leq 0$ if and only if $\frac{m_i t_i}{S_i+1} - 1 \leq 0$, i.e., if and only if $S_i \geq m_i t_i - 1$. In other words, $\beta_i(S_i)$ is concave for $S_i \geq m_i t_i - 1$. Because of the integrality and nonnegativity of S_i, the condition $S_i \geq m_i t_i - 1$ is equivalent to $S_i \geq \max\{\lceil m_i t_i - 1 \rceil, 0\}$. □

The amount $m_i t_i$ represents the average number of parts in the repair pipeline. If this average pipeline stock is smaller than or equal to 1, then $\max\{\lceil m_i t_i - 1 \rceil, 0\} = 0$ and thus $\beta_i(S_i)$ is concave on its whole domain. If the average pipeline stock is larger than 1 (for slow moving SKU's this will not happen in general), then $\max\{\lceil m_i t_i - 1 \rceil, 0\} > 0$ and we exclude solutions $\mathbf{S}$ with $S_i \leq \max\{\lceil m_i t_i - 1 \rceil, 0\}$ from the solution space. The excluded solutions have a very low value for $\beta_i(S_i)$ and they are generally not relevant for problem instances with a high target for the aggregate fill rate.

We now reformulate Problem (Q) as defined in Sect. 2.4. First, we replace the minimization of $EBO(\mathbf{S})$ by the maximization of $\beta(\mathbf{S})$. Second, we limit the solution space to $\mathscr{S}' = \mathscr{S}'_1 \times \mathscr{S}'_2 \times \ldots \times \mathscr{S}'_{|I|}$, with $\mathscr{S}'_i = \{S_i \in \mathbb{N}_0 : S_i \geq m_i t_i - 1\}$, $i \in I$. Hence, we obtain the Problem (Q′):

$$
\begin{array}{lll}
(\mathrm{Q}') & \min & C(\mathbf{S}) \\
 & \max & \beta(\mathbf{S}) \\
 & \multicolumn{2}{l}{\text{subject to } \mathbf{S} \in \mathscr{S}'.}
\end{array}
$$

Problem (Q′) is still separable and the functions $\beta_i(S_i)$ are now increasing and concave on their whole domains $\mathscr{S}'_i$. Hence a set of efficient solutions can be generated by a greedy algorithm. A first efficient solution $\mathbf{S} = (S_1, \ldots, S_{|I|})$ is obtained by setting $S_i = \max\{\lceil m_i t_i - 1 \rceil, 0\}$ for each SKU $i \in I$. This solution is efficient because it has the lowest possible investment. Next, for each SKU i, we compute the increase in $\beta(\mathbf{S})$ relative to the increase in $C(\mathbf{S})$ when S_i would be increased by one unit. The increase in $C(\mathbf{S})$ equals c_i^{a}, while the increase in $\beta(\mathbf{S})$ equals (use (2.12))

$$
\Delta_i\beta(\mathbf{S}) = \frac{m_i}{M}\Delta\beta_i(S_i) = \frac{m_i}{M}\left(\beta_i(S_i+1) - \beta_i(S_i)\right) = \frac{m_i}{M}P\{X_i = S_i\}.
$$

The increase in $\beta(\mathbf{S})$ divided by the increase in $C(\mathbf{S})$ is denoted by

$$
\Gamma_i = \frac{m_i P\{X_i = S_i\}}{M c_i^{\mathrm{a}}}.
$$

The SKU with the highest value for Γ_i is selected, and the corresponding basestock level is increased by one unit (ties are broken with equal probabilities). The new solution S is also efficient and is added to a set of efficient solutions. The generation of efficient solutions is continued until a given aggregate fill rate or inventory investment has been reached, or until some other stop criterium is met. The formal procedure is described in Algorithm 2.2.

Algorithm 2.2 *(Greedy algorithm)*

Step 1 Set $S_i := \max\{\lceil m_i t_i - 1 \rceil, 0\}$ for all $i \in I$;
$S = (S_1, S_2, \ldots, S_{|I|})$;
$\mathscr{E} := \{\mathbf{S}\}$;
Compute $C(\mathbf{S})$ and $\beta(\mathbf{S})$.

Step 2 $\Gamma_i := (m_i P\{X_i = S_i\})/(M c_i^{\mathrm{a}})$ for all $i \in I$;
$k := \arg\max\{\Gamma_i : i \in I\}$;
$\mathbf{S} := \mathbf{S} + \mathbf{e}_k$
$\mathscr{E} := \mathscr{E} \cup \{\mathbf{S}\}$.

Step 3 Compute $C(\mathbf{S})$ and $\beta(\mathbf{S})$;
If 'stop criterium', then stop, else goto Step 2.

For the computation of the pipeline stock probabilities $P\{X_i = S_i\}$ in this algorithm, we advice to use the recursive expression (2.8) for the sake of efficiency. In the following lemma, it is formally stated that Algorithm 2.2 generates efficient solutions for Problem (Q′) (again, the proof follows directly from Theorem 2 in Fox [8]).

Lemma 2.5. *At termination of Algorithm 2.2, the set $\mathscr{E}$ consists of efficient solutions for Problem (Q′).*

Example 2.1 (continued). If we apply the greedy algorithm to our example with $\beta^{\mathrm{obj}} = 0.98$, we find the efficient solutions that are displayed in Table 2.5. After twelve iterations we obtain basestock levels that fulfill the target aggregate fill rate at an inventory investment of 41,000 Euros. In Fig. 2.8, the efficient solutions obtained by both enumeration and the greedy algorithm are displayed. As you can see the greedy algorithm again generates a subset of all efficient solutions.

2.7.5 Aggregate Mean Number of Stockouts

Another service measure is the aggregate mean number of stockouts. This measure counts the number of stockouts for all spare parts together. The mean number of stockouts of SKU i is calculated by the following formula:

$$\alpha_i(S_i) = m_i(1 - \beta_i(S_i)).$$

The aggregate mean number of stockouts is then the sum of the individual mean numbers of stockouts:

$$\alpha(\mathbf{S}) = \sum_{i \in I} \alpha_i(S_i).$$

The increase in $\alpha_i(S_i)$ by increasing the basestock level of SKU i by 1 is:

$$\Delta\alpha_i(S_i) = \alpha_i(S_i + 1) - \alpha_i(S_i) = -m_i P\{X_i = S_i\} \le 0, \qquad S_i \in \mathbb{N}_0.$$

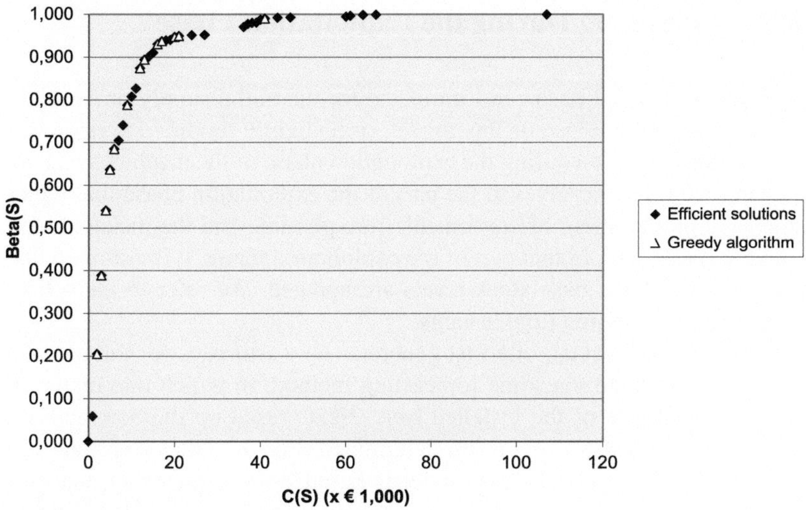

Fig. 2.8 Outcome of the greedy algorithm

Table 2.5 Steps of the greedy algorithm

Iteration	Γ_1	Γ_2	Γ_3	k	S_1	S_2	S_3	$\beta(\mathbf{S})$	$C(\mathbf{S})$ (Euros)
0	–	–	–	–	2	0	0	0.205	2,000
1	$1.83 \cdot 10^{-4}$	$3.45 \cdot 10^{-5}$	$2.02 \cdot 10^{-6}$	1	3	0	0	0.388	3,000
2	$1.53 \cdot 10^{-4}$	$3.45 \cdot 10^{-5}$	$2.02 \cdot 10^{-6}$	1	4	0	0	0.541	4,000
3	$9.54 \cdot 10^{-5}$	$3.45 \cdot 10^{-5}$	$2.02 \cdot 10^{-6}$	1	5	0	0	0.637	5,000
4	$4.77 \cdot 10^{-5}$	$3.45 \cdot 10^{-5}$	$2.02 \cdot 10^{-6}$	1	6	0	0	0.684	6,000
5	$1.99 \cdot 10^{-5}$	$3.45 \cdot 10^{-5}$	$2.02 \cdot 10^{-6}$	2	6	1	0	0.788	9,000
6	$1.99 \cdot 10^{-5}$	$2.87 \cdot 10^{-5}$	$2.02 \cdot 10^{-6}$	2	6	2	0	0.874	12,000
7	$1.99 \cdot 10^{-5}$	$1.20 \cdot 10^{-5}$	$2.02 \cdot 10^{-6}$	1	7	2	0	0.894	13,000
8	$7.10 \cdot 10^{-6}$	$1.20 \cdot 10^{-5}$	$2.02 \cdot 10^{-6}$	2	7	3	0	0.930	16,000
9	$7.10 \cdot 10^{-6}$	$3.33 \cdot 10^{-6}$	$2.02 \cdot 10^{-6}$	1	8	3	0	0.937	17,000
10	$2.22 \cdot 10^{-6}$	$3.33 \cdot 10^{-6}$	$2.02 \cdot 10^{-6}$	2	8	4	0	0.947	20,000
11	$2.22 \cdot 10^{-6}$	$6.93 \cdot 10^{-7}$	$2.02 \cdot 10^{-6}$	1	9	4	0	0.949	21,000
12	$6.16 \cdot 10^{-7}$	$6.93 \cdot 10^{-7}$	$2.02 \cdot 10^{-6}$	3	9	4	1	0.989	41,000

Lemma 2.4 states that $\beta_i(S_i)$ is increasing on its whole domain and concave for $S_i \geq \max\{\lceil m_i t_i - 1 \rceil, 0\}$. Applying Lemma 2.4 to the situation of $\alpha_i(S_i)$ results in the conclusion that $\alpha_i(S_i)$ is decreasing on its whole domain and convex for $S_i \geq \max\{\lceil m_i t_i - 1 \rceil, 0\}$. This can be taken into account for the starting solution in the greedy algorithm. In each iteration of the greedy algorithm, the ratio of the decrease in $\alpha(\mathbf{S})$ and the increase in $C(\mathbf{S})$, due to an increase of the basestock level of SKU i with 1, is measured by Γ_i, which equals:

$$\Gamma_i = \frac{m_i P\{X_i = S_i\}}{c_i^{\mathrm{a}}}.$$

In each iteration, we increase the basestock level of the SKU with the highest Γ_i.

2.8 Inventory Planning During the Exploitation Phase

The basic model of Sect. 2.2 has been formulated for the initial supply problem. In this section, we consider the use of this model, or a slightly modified version, for the tactical inventory planning during the exploitation phase of the machines that are supported. Here, we limit ourselves to the part of the exploitation phase that repair of repairables and procurement of consumables are possible and the installed base has a reasonably stable size. In that part of the exploitation phase, it is common that every month or quarter, the base stock levels are updated. We refer to these time instants as the tactical planning time instants.

At a tactical planning instant, one first generates new estimates for the demand rates m_i, $i \in I$. This is done via some forecasting method, in which one may also incorporate a changed size of the installed base. Next, based on the new demand rates, new base stock levels S_i, $i \in I$, are determined. For the latter step, one can use the basic model in Sect. 2.2. This model is based on a steady-state analysis, and looking at such an analysis is fine only if changes in the basestock levels are effectuated quickly enough for all SKU's. If one orders additional parts for some repairable SKU's and the delivery leadtimes would be 1 year (notice that these lead-times are other times than the repair leadtimes), then one should also look at the transient behavior of these SKU's in the first year. This possible complication is ignored below.

When applying the model in Sect. 2.2 during the exploitation phase, one has already parts on stock or in the repair pipeline, and these amounts have to be taken into account. Let the current level for the inventory position of SKU i be denoted by S_i^{cur}, $i \in I$. The effect of these current base stock levels S_i^{cur} depends on what one can do with the parts of SKU's for which we would like to decrease the base stock level. We distinguish two main cases.

The first case applies for the stock at a local warehouse. Parts that are not needed anymore at the local warehouse go back to the central depot, and the central depot may be able to use those parts for other local warehouses. One can then assume that decreasing the base stock level of an SKU i from S_i^{cur} to a lower level S_i reduces the inventory investment by c_i^{a} per unit decrease. And, increasing the base stock level of an SKU i from S_i^{cur} to a higher level S_i increases the inventory investment by c_i^{a} per unit increase. In other words, the marginal costs for increasing or decreasing base stock levels are the same as for the original problem of Sect. 2.2. Hence, one can optimize the new base stock levels S_i in the same way as before, i.e., as if there is no current stock.

The second case generally applies for the stock at a central depot or the stock in a network as a whole. A decrease of the basestock level for a consumable SKU i means that one part less is procured in the near future. This saves an amount c_i^{a}. Sometimes it is possible to deliver parts back to the original supplier, but one may then receive a lower price back than the original price c_i^{a} for which the part was bought. A decrease of the basestock level for a repairable SKU i may lead to less repairs in the near future or the excess stock is sold, e.g., on a second-hand parts market. In both cases one will save less than c_i^{a} per unit decrease. To capture all cases, we introduce a parameter $\hat{c}_i^{\mathrm{a}}$ ($0 \le \hat{c}_i^{\mathrm{a}} \le c_i^{\mathrm{a}}$) that denotes how much is saved per unit decrease of the basestock level of SKU i, $i \in I$. For an SKU i, the current stock represents a value of $\hat{c}_i^{\mathrm{a}} S_i^{\mathrm{cur}}$, and the cost of changing the basestock level from S_i^{cur} to S_i is

$$\begin{cases} c_i^{\mathrm{a}}(S_i - S_i^{\mathrm{cur}}) & \text{if } S_i \ge S_i^{\mathrm{cur}} \\ \hat{c}_i^{\mathrm{a}}(S_i - S_i^{\mathrm{cur}}) & \text{if } 0 \le S_i < S_i^{\mathrm{cur}}. \end{cases}$$

Adding a constant cost factor $\hat{c}_i^{\mathrm{a}} S_i^{\mathrm{cur}}$ leads to the following modified cost function $C_i(S_i)$:

$$C_i(S_i) = \begin{cases} \hat{c}_i^{\mathrm{a}} S_i & \text{if } 0 \le S_i < S_i^{\mathrm{cur}} \\ \hat{c}_i^{\mathrm{a}} S_i^{\mathrm{cur}} + c_i^{\mathrm{a}}(S_i - S_i^{\mathrm{cur}}) & \text{if } S_i \ge S_i^{\mathrm{cur}}. \end{cases}$$

For the rest, Problem (P) and Problem (Q) remain the same. The modified cost functions $C_i(S_i)$ are increasing and convex. Hence, generating efficient solutions for $C(\mathbf{S})$ and $EBO(\mathbf{S})$ can still be done by the same greedy algorithm (cf. Fox [8], Sect. 8); the factors Γ_i in Algorithm 2.1 are now computed as

$$\Gamma_i = \frac{-\Delta EBO_i(S_i)}{\Delta C_i(S_i)} = \begin{cases} (1 - \sum_{x=0}^{S_i} P\{X_i = x\})/\hat{c}_i^{\mathrm{a}} & \text{if } \hat{c}_i^{\mathrm{a}} > 0 \text{ and } 0 \le S_i < S_i^{\mathrm{cur}} \\ \infty & \text{if } \hat{c}_i^{\mathrm{a}} = 0 \text{ and } 0 \le S_i < S_i^{\mathrm{cur}} \\ (1 - \sum_{x=0}^{S_i} P\{X_i = x\})/c_i^{\mathrm{a}} & \text{if } S_i \ge S_i^{\mathrm{cur}}. \end{cases}$$

If $\hat{c}_i^{\mathrm{a}} = 0$ for some SKU i, then one could say that the first S_i^{cur} are free for this SKU and they should we selected first. We enforce this by saying that $\Gamma_i = \infty$ as long as $S_i < S_i^{\mathrm{cur}}$. The alternative would be to start with $S_i := S_i^{\mathrm{cur}}$ instead of $S_i := 0$ for such an SKU.

2.9 Emergency Shipments

In several practical situations, demand will not be backordered in case of a stockout, but an emergency option will be applied. For example, when our single-location model is applied to a local warehouse, it may be so that a demand is satisfied by an emergency shipment from a central depot or from another local warehouse (in the latter case, the shipment is generally denoted as a lateral transshipment). When our model is applied for a central depot, such an emergency shipment may be possible from an external supplier or another source. Or, for repairables, it may be possible to execute an emergency repair for a part that is already at the repair shop, followed by a fast form of transport. For our spare parts stockpoint, it means that the demand is lost (or a 'lost sale') instead of backordered when it cannot be satisfied from stock. Emergency shipments (or repairs) will be expensive in general, but, in case downtime costs of machines are high, it is natural to apply them. The application of emergency shipments has several consequences for our model description and its whole analysis.

Assume that the average time for an emergency shipment is equal to t_i^{em} for SKU i. Define $W_i(S_i)$ as the mean waiting time for an arbitrary demand for SKU i. It holds that

$$W_i(S_i) = (1 - \beta_i(S_i))t_i^{\text{em}},$$

where $\beta_i(S_i)$ is the fill rate for SKU i. Next, we define $W(\mathbf{S})$ as the mean waiting time for an arbitrary demand for all SKU's together, i.e., the aggregate mean waiting time:

$$W(\mathbf{S}) = \sum_{i \in I} \frac{m_i}{M} W_i(S_i).$$

Then the total downtime of all machines together is equal to $MW(\mathbf{S})$, and the average availability of the machines may be approximated by

$$A(\mathbf{S}) \approx 1 - \frac{MW(S)}{Z}.$$

Hence, setting a minimum level A^{obj} for $A(\mathbf{S})$ is equivalent to setting a maximum level $W^{\text{obj}} = Z(1 - A^{\text{obj}})/M$ for $W(\mathbf{S})$.

With respect to costs, one has two types of costs now. As before, one has costs for buying spare parts at the beginning, with price c_i^{a} for SKU i. In addition, one has costs for the emergency shipments. Each time that one applies an emergency shipment for SKU i, one has costs c_i^{em}. We assume that c_i^{em} contains the costs for a fast delivery from another location. When an emergency shipment is applied, one has one normal replenishment less at our stockpoint, and therefore those costs should be subtracted. The average costs per time unit for SKU i for emergency shipments are equal to $m_i(1 - \beta_i(S_i))c_i^{\text{em}}$. The total costs consist of these costs per time unit and the one-time costs at the beginning. The latter costs may be transformed into inventory holding costs per time unit. We use c_i^{h} to denote the inventory holding cost per time unit per part of SKU i. Then the average costs per time unit for SKU i are equal to

$$\hat{C}_i(S_i) = c_i^{\text{h}} S_i + m_i(1 - \beta_i(S_i))c_i^{\text{em}}, \tag{2.14}$$

and the total average costs are equal to $\hat{C}(\mathbf{S}) = \sum_{i \in I} \hat{C}_i(S_i)$. The optimization problem that we want to solve is as follows:

$$\begin{aligned} (\text{P}'') \qquad \min \qquad & \hat{C}(\mathbf{S}) \\ \text{subject to } & W(\mathbf{S}) \le W^{\text{obj}}, \\ & \mathbf{S} \in \mathscr{S}. \end{aligned}$$

The related multi-objective problem is:

$$\begin{aligned} (\text{Q}'') \qquad \min \qquad & \hat{C}(\mathbf{S}) \\ \min \qquad & W(\mathbf{S}) \\ \text{subject to } & \mathbf{S} \in \mathscr{S}. \end{aligned}$$

The evaluation of a given policy basestock policy $\mathbf{S}$ can still be made per SKU. Under the application of emergency shipments, the number of parts in the repair pipeline of SKU i is bounded from above by S_i. I.e., the behavior of the number of parts in repair of SKU i is no longer as in an $M|G|\infty$ queue but as in an $M|G|c|c$ queue with $c = S_i$ parallel servers, arrival rate m_i, and mean service time t_i. The $M|G|c|c$ queue is also called an *Erlang loss system*. The fill rate $\beta_i(S_i)$ for SKU i is obtained via the so-called Erlang loss probability. The fill rate is equal to the fraction of time that there is at least one part on stock. This is equal to the fraction of time that at least one server is free in the corresponding Erlang loss system. The latter probability is equal to 1 minus the fraction of time that all servers are occupied, i.e., to 1 minus the Erlang loss probability. Hence,

$$\beta_i(S_i) = 1 - \frac{\frac{1}{S_i!}\rho_i^{S_i}}{\sum_{j=0}^{S_i} \frac{1}{j!}\rho_i^j}, \tag{2.15}$$

where $\rho_i := m_i t_i$.

Karush [9] has shown that the Erlang loss probability is strictly convex and decreasing as a function of the number of servers (see also Remark 2 in [11]). This implies that $\beta_i(S_i)$ is strictly concave and increasing on its whole domain. As a result:

- For each $i \in I$, $W_i(S_i)$ is decreasing and concave on its whole domain.
- For each $i \in I$, $\hat{C}_i(S_i)$ is convex on its whole domain. The function $\hat{C}_i(S_i)$ may now be decreasing for smaller values of S_i, because of the presence of the emergency costs.

Let $S_{i,\min} := \arg\min \hat{C}_i(S_i)$. Then, obviously, for Problem (P$''$) and its corresponding multi-objective programming problem (Q$''$), we may exclude solutions with $S_i < S_{i,\min}$ for some $i \in I$. Then, one can generate efficient solutions for $C(\mathbf{S})$ and $W(\mathbf{S})$ in a similar way as for Problem (Q) in Sect. 2.4 (see [8], Sect. 8); the factors Γ_i are now computed as $\Gamma_i := -\Delta_i W(\mathbf{S})/\Delta_i \hat{C}(\mathbf{S})$ with $\Delta_i W(\mathbf{S}) = (m_i/M)\Delta W_i(S_i) = (m_i/M)(W_i(S_i+1) - W_i(S_i))$ and $\Delta_i \hat{C}(\mathbf{S}) = \Delta \hat{C}_i(S_i) = \hat{C}_i(S_i+1) - \hat{C}_i(S_i)$, $S_i \geq S_{i,\min}$. This results in the following algorithm.

Algorithm 2.3 *(Greedy algorithm)*

Step 1 $\quad S_{i,\min} := \arg\min \hat{C}_i(S_i)$ for all $i \in I$;
  Set $S_i := S_{i,\min}$ for all $i \in I$, and $\mathbf{S} = (S_{1,\min}, \ldots, S_{|I|,\min})$;
  $\mathscr{E} := \{\mathbf{S}\}$;
  Compute $\hat{C}(\mathbf{S})$ and $W(\mathbf{S})$.
Step 2 $\quad \Gamma_i := -(m_i \Delta W_i(S_i))/(M \Delta \hat{C}_i(S_i))$ for all $i \in I$;
  $k := \arg\max\{\Gamma_i : i \in I\}$;
  $\mathbf{S} := \mathbf{S} + \mathbf{e}_k$;
  $\mathscr{E} := \mathscr{E} \cup \{\mathbf{S}\}$.
Step 3 $\quad$ Compute $\hat{C}(\mathbf{S})$ and $W(\mathbf{S})$;
  If 'stop criterium', then stop, else goto Step 2.

In the following lemma, it is formally stated that Algorithm 2.3 generates efficient solutions for Problem (Q″). The proof of this lemma follows from Fox [8] (Sect. 8).

Lemma 2.6. *At termination of Algorithm 2.3, the set $\mathscr{E}$ consists of efficient solutions for Problem (Q″).*

2.10 Extensions

In this section we describe a few extensions to our model.

2.10.1 Consumables and Condemnation

In Sect. 2.2, we use the terminology that is common for repairable spare parts. Nevertheless, the model also applies when all SKU's, or some of the SKU's, are consumable. For a consumable, we assume that a new part is procured each time that a demand occurs, and we assume i.i.d. procurement leadtimes, which again are denoted by t_i.

One can also have a kind of mixture of the situations for a repairable and a consumable, respectively. So far, for a repairable, we assume that a repair is always executed and it is always successful. In practice, this is often more subtle. In many cases, components can fail due to various reasons. Some of the resulting defects may be repairable and others not. It may also be that a part is only repaired for a limited number of times, because its performance slowly decreases after each repair. One generally refers to these phenomena as *condemnation*. From a modeling point of view, condemnation can be easily incorporated, which was already noticed by [5]. The idea is to introduce a parameter r_i that represents the probability that a failed part of SKU i can be repaired. Next, we distinguish a mean repair leadtime t_i^{rep} and a mean procurement leadtime t_i^{proc} (the procurement leadtime for SKU i may be stochastic as long as realizations of the leadtime for different order are independent). Then an arbitrary failed part leads to the arrival of a ready-for-use/new part at the most upstream location after an average leadtime $t_i = r_i t_i^{\text{rep}} + (1 - r_i) t_i^{\text{proc}}$. These t_i's are the leadtimes that can be used in the basic model without emergency shipments, and similarly in the model with emergency shipments.

2.10.2 Excluding Pipeline Stock

In Sect. 2.10.1, we denoted that the model of Sect. 2.2 also applies for consumables. In that basic model, we look at inventory investment. In Sect. 2.9, we describe the model with emergency shipments, and we have switched to inventory holding costs so that these costs can be added to the emergency shipments costs to obtain the average costs per time unit. When determining inventory holding costs for consumables that are procured at an external supplier, it is more appropriate to exclude the pipeline stock (i.e., the parts that are on order but not on hand yet). For a consumable SKU i in the model of Sect. 2.9, replenishment orders for single units are placed with rate $m_i\beta_i(S_i)$, and, by Little's law, the average pipeline stock is $m_i\beta_i(S_i)t_i$. This leads to the following modified expression for the average costs per time unit for SKU i, as given in (2.14):

$$\begin{aligned}\hat{C}_i(S_i) &= c_i^{\mathrm{h}}(S_i - m_i\beta_i(S_i)t_i) + m_i(1-\beta_i(S_i))c_i^{\mathrm{em}} \\ &= c_i^{\mathrm{h}}(S_i - m_it_i) + m_i(1-\beta_i(S_i))(c_i^{\mathrm{em}} + c_i^{\mathrm{h}}t_i).\end{aligned}$$

This function is still convex (like the function $\hat{C}_i(S_i)$ in (2.14)), and thus one can still follow the same solution procedure as in Sect. 2.9.

2.10.3 Batching

In our basic model, we assume one-for-one replenishments for all SKU's. For a local warehouse, this is generally justified because it receives consolidated replenishments for all SKU's together from a central depot. However, at a central depot, one may send failed parts into repair at external repair shops or one orders consumables at outside suppliers, and then some form of batching may be desired. Reasons for using batching may be fixed setup costs for certain repair activities, fixed ordering and delivery costs that are charged by external suppliers, or pack sizes that are prescribed by suppliers. Applying the logic of the EOQ rule shows that generally one-for-one replenishments will make sense for the more expensive SKU's, which have high inventory holding costs and/or low demand rates. For less expensive components, however, it may be appropriate to use a fixed batch size Q, and thus to follow an (s,Q)-policy instead of a basestock policy. Consider the basic model (without emergency shipments), but assume now that for each SKU i, a batch of Q_i failed parts is sent into repair as soon as the inventory position of SKU i drops to its reorder level s_i (≥ -1). (Notice that a basestock policy with basestock level $S_i = s_i + 1$ is obtained in case $Q_i = 1$.) The repair leadtime for such a batch is always equal to t_i (we now assume that the repair leadtime is deterministic). The performance for SKU i is then obtained by making use of the following two key properties:

- The inventory level $OH_i(t) - BO_i(t)$ of SKU i at an arbitrary time point t is given by the inventory position at time point $t - t_i$ minus the demand in the time interval $[t - t_i, t)$ (see Sect. 5.3.2 of Axsäter [1]).
- The inventory position at time point $t - t_i$ is as the inventory position at an arbitrary time point and thus is uniformly distributed on the integers $s_i + 1, s_i + 2, \ldots, s_i + Q_i$ (see Proposition 5.1 of [1]).

The demand during time interval $[t-t_i,t)$ is Poisson distributed with mean $m_i t_i$, and thus has the same distribution as the pipeline stock X_i in the basic model. Hence, $OH_i(t)-BO_i(t)$ is equal to $s_i+U_i-X_i$, where U_i is a uniformly distributed random variable on $\{1,\ldots,Q_i\}$. For the on-hand stock and the number of backorders in steady state, we then obtain:

$$OH_i = (s_i+U_i-X_i)^+,$$
$$BO_i = (X_i-(s_i+U_i))^+.$$

These expressions generalize the expressions (2.4)–(2.5) of the basic model. The rest of the analysis goes along similar lines as for the basic model. In particular, the mean number of backorders is now denoted by $EBO_i(s_i)$ and may be shown to be decreasing and convex for $s_i \geq -1$.

For the model with emergency shipments, it is less easy to incorporate a fixed batch size. Emergency shipments lead to lost sales for the inventory of spare parts, and, generally spoken, lost sales models are much harder to analyze than backordering models (see also Bijvank and Vis [2]). The two key properties as described above for the backordering case, do not hold for the lost sales case. Hence, for the lost sales case, one has to rely on approximate evaluation methods; see [2] for further references.

2.10.4 Criticality

So far, we have assumed that all components are equally critical, i.e., a delay of x hours/days in fulfilling a spare parts demand is equally bad for all SKU's. When considering system availability, we assume that each component is critical. That is, an entire system goes down when a component has failed and can not be immediately replaced by a spare part. The reality in practice is often more sophisticated. The criticality of a component is related to the consequences for the system and system output if that component is not replaced immediately. These consequences may depend on the failure mode of a failed part, the position of a component in the system (an SKU may occur at multiple places in a system, and the criticality may differ per position), the level of redundancy per position, and so on. A good way to address these factors is to go back to the reliability data of a system and to incorporate the above factors in the modeling. This has been done by Van Jaarsveld and Dekker [21].

Generally, it appears to be difficult to quantify the criticality of all components. But, suppose that this has been done. Then it is possible to classify components in multiple criticality classes, and next a service target may be specified per class or one adds weights and optimizes under weighted service level constraints.

2.11 Concluding Remarks

A single-item version of the basic model in this chapter was formulated for the first time in 1966 by Feeney and Sherbrooke [5], who also discussed extensions to compound Poisson demand processes and the emergency shipments case (also denoted as the lost sales case). Shortly later, Sherbrooke [17] extended this model to a multi-item distribution system with one central warehouse and multiple local warehouse. This was the so-called METRIC model, which we will discuss extensively in Chap. 6. For a heuristic optimization of basestock levels, a so-called marginal analysis was introduced, which is like the greedy algorithm as formulated in this chapter. The paper by Sherbrooke led to a big stream of papers on spare parts models.

A better understanding for the marginal analysis of Sherbrooke [17] and the quality of its solutions was developed later; see e.g. Sherbrooke [18] and Wong et al. [22]. In Sect. 2.12 of Sherbrooke [18], a justification for the marginal analysis is given for the single-location model. However, a link with efficient solutions for a corresponding multi-objective programming problem and a real proof that it leads to optimal solutions for specific target values for the aggregate mean number of backorders (cf. Sect. 2.4.2) are absent in those works. To the best of our knowledge, that link and proof were given for the first time in Van Houtum and Hoen [20].

Differences between the system and item approach, as presented in Sect. 2.6, were studied by multiple authors; see e.g. Rustenburg et al. [16] and Thonemann et al. [19]. In the latter paper, it has been shown that cost differences between the SKU's are the main factor to lead to large cost differences between the system and item approach.

Problems

2.1. Consider a single warehouse for which the basic model of Sect. 2.2 applies. We have $|I| = 3$ SKU's. The data for the SKU's are as follows:

i	m_i (per month)	t_i (months)	c_i^{a} (Euros)
1	1.0	1.0	500
2	0.4	1.5	1,400
3	0.2	2.0	4,000

The maximum level for the aggregate mean number of backorders is 0.2.

(a) Compute the probabilities $P\{X_i = x\}$ and $P\{X_i \leq x\}$ for $i = 1,2,3$ and $0 \leq x \leq 6$. Hint: Use the recursion in (2.8) for the computation of the probabilities $P\{X_i = x\}$.

(b) Apply Algorithm 2.1 to generate efficient solutions for Problem (Q). (Use an appropriate stop criterium.)
(c) What is the first solution $\mathbf{S}$ generated under (b) for which the aggregate mean number of backorders is at most 0.2? What are the average costs under this solution?

2.2. Consider Problem 2.1, and suppose that the item approach with $EBO^{\text{obj}} = 0.2$ is applied to generate a feasible solution. Denote this solution as $\mathbf{S}^{\text{item}}$.

(a) Determine $\mathbf{S}^{\text{item}}$, $EBO(\mathbf{S}^{\text{item}})$, and $C(\mathbf{S}^{\text{item}})$.
(b) What is the costs difference with the solution obtained under part (c) of Problem 2.1?
(c) Consider the efficient solutions obtained via the system approach, i.e., via Algorithm 2.1; see also part (b) of Problem 2.1. What is the cheapest solution $\mathbf{S}$ among them with $EBO(\mathbf{S}) \leq EBO(\mathbf{S}^{\text{item}})$, and how large is the costs difference? What is the solution $\mathbf{S}$ among them with the lowest aggregate mean number of backorders and costs $C(\mathbf{S}) \leq C(\mathbf{S}^{\text{item}})$, and how large is the difference in the aggregate mean number of backorders when comparing this solution $\mathbf{S}$ to $\mathbf{S}^{\text{item}}$?

2.3. Consider Problem 2.1. We are interested in additional performance measures for the solution $\mathbf{S}$ obtained under part (c) of that problem.

(a) Determine the aggregate mean waiting time $W(\mathbf{S})$ under solution $\mathbf{S}$.
(b) What is the aggregate fill rate $\beta(\mathbf{S})$ under solution $\mathbf{S}$?
(c) A fraction $1 - \beta(\mathbf{S})$ of all demands is backordered, and they are satisfied after a certain delay. What is the average delay for an arbitrary backordered demand?

2.4. Consider the problem as described in Sect. 2.7.3. First, prove that

$$\Delta^2 PBO_i(S_i) = \left(1 - \frac{m_i t_i}{S_i + 2}\right) P\{X_i = S_i + 1\}, \qquad S_i \in \mathbb{N}_0.$$

Next, use this equation to prove that $PBO_i(S_i)$ is convex for $S_i \geq \max\{\lceil m_i t_i - 2 \rceil, 0\}$.

2.5. Consider Problem 2.1, but suppose that one has already certain parts on stock: 1 part of SKU 1, 2 parts of SKU 2, and 3 parts of SKU 3. The parts are bought from an external supplier and it is not possible to sell parts back to the supplier. The requirement for the aggregate mean number of backorders remains the same.

(a) Problem (Q) as described for the standard problem of Sect. 2.2 becomes slightly different. Give the adapted problem formulation.
(b) Formulate a greedy algorithm that gives efficient solutions for the adapted Problem (Q).
(c) Apply the algorithm formulated under (b).

(d) What is the first solution $\mathbf{S}$ generated under (c) for which the aggregate mean number of backorders is at most 0.2? What are the average costs under this solution?

2.6. Consider Problem 2.5, but now assume that parts can be sold back to the supplier for 50 % of the selling price. Answer the same questions as for Problem 2.5.

2.7. Consider the extension of the basic model to the case with emergency shipments as described in Sect. 2.9. In the analysis of this extended model, one uses the property that the Erlang loss probability is strictly convex and decreasing as a function of the number of servers. For an Erlang loss system with $c \in \mathbb{N}_0$ servers and offered load $\rho > 0$, the Erlang loss probability is given by:

$$L(c,\rho) = \frac{\rho^c/c!}{\sum_{x=0}^{c} \rho^x/x!}.$$

Prove that $L(c,\rho)$ is strictly decreasing as a function of $c \in \mathbb{N}_0$.

2.8. Consider the problem of Sect. 2.9. Assume that the replenishment leadtimes are exponentially distributed (i.e., we consider a special case). Determine the steady-state distribution for the number of parts on order of an SKU i via a Markov analysis. Verify the correctness of (2.15) for this special case.

2.9. Consider the generalization of the basic model to the use of given batch sizes for the inventory control of all SKU's; see Sect. 2.10.3. For each SKU $i \in I$, one then follows an (s_i, Q_i)-policy, where Q_i represents the given batch size for SKU i and s_i (≥ -1) is the reorder level. The mean number of backorders is denoted by $EBO_i(s_i)$. Prove that $EBO_i(s_i)$ is decreasing and convex for $s_i \geq -1$.

2.10. Consider a company with its own maintenance department and a own spare parts stock. The spare parts are used for corrective maintenance at a group of machines. For the inventory control, the company wants to use a system approach. The service measure that they work with is the aggregate fill rate, cf. Sect. 2.7.4.

To learn more about how the system approach works, the company wants to consider a set of three representative spare parts. The data for these parts are as follows:

$$m_1 = 15,\ m_2 = 5,\ m_3 = 1 \ \text{ (in demands per year)},$$
$$t_1 = t_2 = t_3 = \frac{1}{6} \ \text{ (in years)},$$
$$c_1^{\text{a}} = 300,\ c_2^{\text{a}} = 800,\ c_3^{\text{a}} = 5{,}000 \ \text{ (in Euros)}.$$

(a) Apply the greedy algorithm of Sect. 2.7.4 to generate efficient solutions for $C(\mathbf{S})$ and $\beta(\mathbf{S})$. Plot these solutions in a figure. List the solutions also in a table. What is now the cheapest solution with an aggregate fill rate of at least $\beta^{\text{obj}} = 0.98$?

The company is not willing to apply a pure form of the system approach. They want to avoid too high basestock levels for cheap parts, because there is always inaccuracy in estimating demand rates and then a too high basestock level may lead to dead stock in the future. For each SKU, they want a fill rate of at most 0.998. In addition, they do not want too low basestock levels for expensive SKU's, because that may hinder acceptance of the use of the system approach. For each SKU, they want a fill rate of at least 0.90.

(b) With these extra constraints, one obtains a variant of Problem (Q′) of Sect. 2.7.4. Formulate this variant.
(c) For this variant of Problem (Q′), one can again formulate a greedy algorithm to generate efficient solutions. Formulate such a greedy algorithm and explain why it will generate efficient solutions.
(d) Apply the greedy algorithm of (c), and plot the generated solutions in the same figure as the solutions obtained under (a). List the solutions also in a table. What is now the cheapest solution with an aggregate fill rate of at least $\beta^{\text{obj}} = 0.98$? How large is the increase in costs in comparison to the solution obtained under (a)?

References

1. Axsäter, S.: Inventory Control, 2nd edn. Springer, New York (2006)
2. Bijvank, M., Vis, I.F.A.: Lost-sales inventory theory: a review. Eur. J. Oper. Res. **215**, 1–13 (2011)
3. Dantzig, G.B., Wolfe, P.: Decomposition principle for linear programs. Oper. Res. **8**, 101–111 (1960)
4. Everett, H. III: Generalized lagrange multiplier method for solving problems of optimum allocation of resources. Oper. Res. **11**, 399–417 (1963)
5. Feeney, G.J., Sherbrooke, C.C.: The $(s-1,s)$ inventory policy under compound Poisson demand. Manag. Sci. **12**, 391–411 (1966)
6. Fisher, M.L.: The Lagrangian relaxation method for solving integer programming problems. Manag. Sci. **27**, 1–18 (1981)
7. Fisher, M.L.: An application oriented guide to Lagrangian relaxation. Interfaces **15**, 10–21 (1985)
8. Fox, B.: Discrete optimization via marginal analysis. Manag. Sci. **13**, 210–216 (1966)
9. Karush, W.: A queueing model for an inventory problem. Oper. Res. **5**, 693–703 (1957)
10. Kranenburg, A.A.: Spare parts inventory control under system availability constraints. Ph.D. thesis, Eindhoven University of Technology (2006). http://w3.tue.nl/en/services/library/digilib/publications_from_tue/dissertations/
11. Kranenburg, A.A., Van Houtum, G.J.: Cost optimization in the $(S-1,S)$ lost sales inventory model with multiple demand classes. OR Lett. **35**, 493–502 (2007)
12. Little, J.D.C.: A proof of the queuing formula $L = \lambda W$. Oper. Res. **9**, 383–387 (1961)
13. Palm, C.: Analysis of the Erlang traffic formula for busy-signal arrangements. Ericsson Technics **4**, 204–212 (1938)

14. Porteus, E.L.: Foundations of Stochastic Inventory Theory. Stanford University Press, Stanford (2002)
15. Rustenburg, W.D.: A system approach to budget-constrained spare parts management. Ph.D. thesis, Eindhoven University of Technology (2000). http://w3.tue.nl/en/services/library/digilib/publications_from_tue/dissertations/
16. Rustenburg, W.D., Van Houtum, G.J., Zijm, W.H.M.: Spare parts management at complex technology-based organizations: an agenda for research. Int. J. Prod. Econ. **71**, 177–193 (2001)
17. Sherbrooke, C.C.: METRIC: a multi-echelon technique for recoverable item control. Oper. Res. **16**, 122–141 (1968)
18. Sherbrooke, C.C.: Optimal Inventory Modeling of Systems: Multi-echelon Techniques. Kluwer Academic, Boston/Dordrecht/London (2004)
19. Thonemann, U.W., Brown, A.O., Hausman, W.H.: Easy quantification of improved spare parts inventory policies. Manag. Sci. **48**, 1213–1225 (2002)
20. Van Houtum, G.J., Hoen, K.M.R.: Single-location, multi-item inventory models for spare parts. Lecture notes, Eindhoven University of Technology (2008)
21. Van Jaarsveld, W., Dekker, R.,: Spare parts stock control for redundant systems using reliability centered maintenance data. Reliab. Eng. Syst. Saf. **96**, 1576–1586 (2011)
22. Wong, H., Kranenburg, A.A., Van Houtum, G.J., Cattrysse, D.: Efficient heuristics for two-echelon spare parts inventory systems with an aggregate mean waiting time constraint per local warehouse. OR Spectr. **29**, 699–722 (2007)

Chapter 3
Multiple Machine Types with Commonality

3.1 Introduction

In Chap. 2, we considered a single location model for the support of an installed base of machines of the same type and with the same target service level. In this and the next chapter, we look at single location models for the support of machines of multiple machine types and/or differentiated target service levels. Here, the multiple machine types are assumed to have common critical components. In this chapter, we focus on machines of multiple machine types, where we implicitly assume that all machines have similar target service levels (so, no differentiation for different machines types or for machines of the same type). In that case, it is natural to satisfy demands of all machine types as long as there is stock available. In the next chapter, we will focus on classes of customers, with identical or different machines, and with significant differences among their target service levels. In that case, one generally wants to differentiate between demands from different classes; e.g., one may want to keep back the last parts on stock for the class(es) with the highest service target(s). We then get more complicated types of inventory policies (so-called critical level policies; see Chap. 4).

For the model presented in this chapter, we focus on the perspective of a local warehouse where emergency shipments are made in case of a stockout. The model extends the model of Sect. 2.9 to multiple machine types (with common components), and we obtain an optimization problem with multiple aggregate mean waiting time constraints. We formulate a greedy heuristic for this optimization problem, and we will show that this heuristic is efficient and also effective for instances with sufficiently many SKU's. The effectiveness is shown via a lower bound obtained by Dantzig-Wolfe decomposition. Via this lower bound, we get a second heuristic that is also considered.

Like in Chap. 2, it is easy to switch to variants of the presented model, including the variant without emergency shipments.

G.J. van Houtum, B. Kranenburg, *Spare Parts Inventory Control under System Availability Constraints*, International Series in Operations Research & Management Science 227, DOI 10.1007/978-1-4899-7609-3_3

In addition to solving the above problem, we also study the benefits of common components of multiple machine types. Without commonality, or without using the commonality, one would keep separate stocks of spare parts per machine type. With commonality, one can use one shared stock, which generally will lead to lower spare parts costs. We study the benefits of commonality for varying levels of commonality.

The organization of this chapter is as follows. We present our model in Sect. 3.2. Next, the greedy heuristic and Dantzig-Wolfe decomposition are described in Sect. 3.3. The performance of the heuristics and the effect of commonality on total costs are analyzed via computational experiments in Sect. 3.4. After that, we present a case study at ASML in Sect. 3.5 and we make concluding remarks in Sect. 3.6.

3.2 Model

Consider a single warehouse where spare parts are kept on stock to support machines of multiple machine types. The supplier of the warehouse is either another warehouse at a higher echelon level of a network or a repair shop, where failed parts are repaired. The machine types have common critical components. We look at all critical components of the machines, and refer to them as stock-keeping units (SKU's). Each SKU may be either a repairable or a consumable; in the rest of the chapter, we use the terminology that is common for repairables, but the whole chapter applies also for consumables.

Let I denote the set of SKU's, with $|I| \geq 1$. The SKU's are numbered as $1,\ldots,|I|$. Let N be the set of machine types, with $|N| \geq 1$; the machine types are numbered as $1,\ldots,|N|$. For each SKU $i \in I$ and machine type $n \in N$, failures (demands) are assumed to occur according to a Poisson process with constant rate $m_{i,n}$ (≥ 0). If SKU i does not occur in the material breakdown structure of machines type n, then $m_{i,n} = 0$ by definition. The total stream of demands for SKU i, $i \in I$, is also a Poisson stream; its rate is $\mu_i := \sum_{n\in N} m_{i,n}$, and we assume that $\mu_i > 0$. The total demand rate for machine type n is denoted by $M_n := \sum_{i\in I} m_{i,n}$, $n \in N$, and we assume that $M_n > 0$.

If one of the parts of a machine fails, the machine is down and the defective part has to be replaced by a spare part. A failure of a machine is always caused by one defective part, and can be remedied by replacing that part only. If a requested part $i \in I$ is available at the warehouse, it is delivered immediately. Otherwise, an emergency shipment from the supplier is made to fulfill the demand. For the stock at the warehouse, such a demand can be considered as a lost sale. The average time for an emergency shipment from the supplier to the warehouse is $t_i^{\mathrm{em}}(\geq 0)$.

For each SKU $i \in I$, the stock in the warehouse is controlled by a base stock policy with base stock level S_i ($\in \mathbb{N}_0$). The inventory holding cost per time unit for one unit of SKU i is $c_i^{\mathrm{h}}(> 0)$. When a part i in the warehouse stock is used to fulfill customer demand, a ready-for-use part i arrives in the warehouse to refill the stock after a regular repair/replenishment leadtime with mean t_i ($> t_i^{\mathrm{em}}$). Per SKU, leadtimes are independent and identically distributed, and lead times for different SKU's are independent.

In addition to the inventory holding costs, we have costs for regular replenishments and emergency shipments. For each demand for an SKU i, we have either a regular replenishment or a more expensive, emergency shipment. For each emergency shipment of SKU i, we only count the *extra* costs, i.e., we count the costs of an emergency shipment minus the costs of a regular replenishment. The extra costs are denoted by the cost factor $c_i^{\mathrm{em}} (\geq 0)$. Notice that in this way a fixed cost factor $\sum_{i \in I} m_i \times$ (costs for one regular shipment of SKU i) is excluded from the model.

If the supplier is another warehouse, an emergency shipment may mean that a faster picking procedure and transport channel is used than for a regular replenishment. If the supplier is a repair shop, then an emergency shipment may mean that one of the parts is repaired (or finished) with the highest possible speed. In both cases, an alternative is that a part is obtained from another source. In that case, we assume that the part that just failed is sent back to this other source (either immediately or after repair), so that the inventory position remains constant.

For each SKU $i \in I$, let $W_i(S_i)$ and $\beta_i(S_i)$, denote the mean waiting time and fill rate, respectively. For each $n \in N$, let $\hat{W}_n(\mathbf{S})$ denote the aggregate mean waiting time, where $\mathbf{S} := (S_1, \ldots, S_{|I|})$ denotes an overall basestock policy. The behavior of the number of parts in the replenishment pipeline of SKU i is as the number of jobs in the Erlang loss model with arrival rate μ_i and mean service time t_i (see also Sect. 2.9). Hence $\beta_i(S_i)$ is equal to one minus the Erlang loss probability (cf. (2.15)):

$$\beta_i(S_i) = 1 - \frac{\frac{1}{S_i!}\rho_i^{S_i}}{\sum_{j=0}^{S_i} \frac{1}{j!}\rho_i^j}, \tag{3.1}$$

where $\rho_i := \mu_i t_i$. Further,

$$W_i(S_i) = (1 - \beta_i(S_i)) t_i^{\mathrm{em}}, \quad i \in I, \tag{3.2}$$

and the aggregate mean waiting times $\hat{W}_n(\mathbf{S})$ are weighted sums of the average waiting times $W_i(S_i)$ for individual SKU's, with the fractions $m_{i,n}/M_n$ as weights:

$$\hat{W}_n(\mathbf{S}) = \sum_{i \in I} \frac{m_{i,n}}{M_n} W_i(S_i), \quad n \in N.$$

The objective for the aggregate mean waiting time for machine type n is denoted by $\hat{W}_n^{\mathrm{obj}}$.

For each SKU i, the inventory holding costs per time unit are $c_i^{\mathrm{h}} S_i$, and the average costs per time unit for emergency shipments are equal to $\mu_i(1 - \beta_i(S_i)) c_i^{\mathrm{em}}$. Hence, the average costs for SKU i are equal to

$$C_i(S_i) := c_i^{\mathrm{h}} S_i + \mu_i(1 - \beta_i(S_i)) c_i^{\mathrm{em}}, \tag{3.3}$$

and the total average costs are equal to $C(\mathbf{S}) = \sum_{i \in I} C_i(S_i)$.

The objective is to minimize the average costs subject to the aggregate mean waiting time constraints for the machine types:

$$\begin{aligned}
\text{(P)} \qquad \min \qquad & C(\mathbf{S}) \\
\text{subject to } & \hat{W}_n(\mathbf{S}) \leq \hat{W}_n^{\text{obj}}, \quad n \in N, \\
& \mathbf{S} \in \mathscr{S},
\end{aligned}$$

where $\mathscr{S} = \{\mathbf{S} = (S_1, \ldots, S_{|I|}) \mid S_i \in \mathbb{N}_0, \forall i \in I\}$. The optimal costs of Problem (P) are denoted by C_{P}. An optimal solution of (P) is denoted by $\mathbf{S}^* = (S_1^*, \ldots, S_{|I|}^*)$.

Notice that the above model describes the situation with shared stock for all machine types together. The situation with a separate stock per machine type can be analyzed by applying the above model for each machine type individually.

Example 3.1. A manufacturer of capital goods keeps spare parts on stock in a single warehouse to support a reasonably large number of installed machines. We distinguish two machine types and three SKU's (i.e., $|N| = 2$ and $|I| = 3$). SKU 1 only occurs in the configuration of machine type 1, and its demand rate is equal to 1.2 demands per month. SKU 2 only occurs in the configuration of machine type 2, and its demand rate is equal to 0.7 demands per month. SKU 3 occurs in both machine types, and its demand rate is equal to 1 demand per month for machine type 1 and 0.7 demands per month for machine type 2. So,

$$\begin{aligned}
& m_{1,1} = 1.2, \;\; m_{1,2} = 0, \;\; m_{2,1} = 0, \;\; m_{2,2} = 0.7, \;\; m_{3,1} = 1, \;\; m_{3,2} = 0.7, \\
& \mu_1 = 1.2, \;\; \mu_2 = 0.7, \;\; \mu_3 = 1.7, \\
& M_1 = 2.2, \;\; M_2 = 1.4.
\end{aligned}$$

All failed parts go to an internal repair shop. The average repair leadtimes are equal to 1 month for all three SKU's ($t_1 = t_2 = t_3 = 1$). The emergency shipment times are equal to 2 days for all three SKU's ($t_1^{\text{em}} = t_2^{\text{em}} = t_3^{\text{em}} = 2$ days). The extra cost for each emergency shipment are 750 Euros ($c_1^{\text{em}} = c_2^{\text{em}} = c_3^{\text{em}} = 750$ Euros). The price of SKU 1 is 10,000 Euros, the price of SKU 2 is 20,000 Euros, and the price of SKU 3 is 7,000 Euros. The inventory holding costs per month are 1.5 % of the price of a new part. Hence, $c_1^{\text{h}} = 0.015 \cdot 10{,}000 = 150$ Euros per part per month, and similarly $c_2^{\text{h}} = 300$ and $c_3^{\text{h}} = 105$ Euros per part per month. For machine type 1 and 2, it is specified that the aggregate mean waiting time may not exceed 0.2 and 0.15 days, respectively ($\hat{W}_1^{\text{obj}} = 0.2$ and $\hat{W}_1^{\text{obj}} = 0.15$ days per demand).

Remark 3.1. In case the average emergency shipment times t_i^{em} are the same for all SKU's, i.e., $t_i^{\text{em}} = t^{\text{em}}$ for all $i \in I$, the aggregate mean waiting time constraints may be rewritten as

$$\sum_{i \in I} \frac{m_{i,n}}{M_n} \beta_i(S_i) \geq \hat{\beta}_n^{\text{obj}}, \qquad n \in N,$$

with

$$\hat{\beta}_n^{\text{obj}} = 1 - \frac{\hat{W}_n^{\text{obj}}}{t^{\text{em}}}.$$

Thus, in that case, the aggregate mean waiting time constraints are equivalent to aggregate fill rate constraints.

3.3 Analysis

For small problem instances of Problem (P), an exact solution can be obtained by enumeration, which is described in Sect. 3.3.1. For large instances, we have to rely on heuristics. In Sect. 3.3.2, we describe an appropriate greedy heuristic. Next, in Sect. 3.3.3, we apply Dantzig-Wolfe decomposition, which leads to both a heuristic solution and a lower bound for the optimal costs C_{P} of Problem (P). Contrary to the exact solution via enumeration, this latter lower bound can also be computed for large instances, and this enables the testing of the two heuristics for large instances (see also Sect. 3.4).

3.3.1 Enumeration

We first limit the solution space from below. As in Sect. 2.9, the item fill rate $\beta_i(S_i)$ given in (3.1) is strictly concave and increasing on its whole domain, and thus $W_i(S_i)$ is strictly convex and decreasing and $C_i(S_i)$ is strictly convex. Let

$$S_{i,\mathrm{lb}} := \min\{S_i \in \mathbb{N}_0 | \Delta C_i(S_i) > 0\}, \quad i \in I, \tag{3.4}$$

where $\Delta C_i(S_i)$ is the first order difference function of $C_i(S_i)$. There exists an optimal solution $\mathbf{S}^* = (S_1^*, \ldots, S_{|I|}^*)$ with $S_i^* \geq S_{i,\mathrm{lb}}$ for all $i \in I$ (suppose $\mathbf{S}^* = (S_1^*, \ldots, S_{|I|}^*)$ is an optimal solution with $S_i^* < S_{i,\mathrm{lb}}$ for some $i \in I$, then $\hat{\mathbf{S}}^* = (\hat{S}_1^*, \ldots, \hat{S}_{|I|}^*)$ with $\hat{S}_i^* = \max\{S_i^*, S_{i,\mathrm{lb}}\}$ for all $i \in I$ is also an optimal solution). So, in the enumeration procedure, we may exclude all solutions with $S_i < S_{i,\mathrm{lb}}$ for some $i \in I$. Further, it holds that the optimal costs are bounded from below by $C_{\mathrm{P}}^{\mathrm{lb}} := \sum_{i \in I} C_i(S_{i,\mathrm{lb}})$.

Next, we limit the solution space from above. For that purpose an upper bound $C_{\mathrm{P}}^{\mathrm{ub}}$ for the optimal costs C_{P} is needed. For this upper bound, we may take the costs of a feasible solution. A straightforward feasible solution is obtained as follows. Let $\hat{W}_{\min}^{\mathrm{obj}} := \min_{n \in N} \hat{W}_n^{\mathrm{obj}}$, and define

$$S_i = \min\{x \geq S_{i,\mathrm{lb}} | W_i(x) \leq \hat{W}_{\min}^{\mathrm{obj}}\}. \tag{3.5}$$

Then $\mathbf{S} = (S_1, \ldots, S_{|I|})$ is a feasible solution, and $C_{\mathrm{P}}^{\mathrm{ub}} = C(\mathbf{S})$ is an upper bound for the optimal costs. The gap between $C_{\mathrm{P}}^{\mathrm{ub}}$ and $C_{\mathrm{P}}^{\mathrm{lb}}$ is an upper bound for how much extra costs can be made for each SKU i in addition to $C_i(S_{i,\mathrm{lb}})$. Define

$$S_{i,\mathrm{ub}} := \max\{x \geq S_{i,\mathrm{lb}} | C_i(x) - C_i(S_{i,\mathrm{lb}}) \leq C_{\mathrm{P}}^{\mathrm{ub}} - C_{\mathrm{P}}^{\mathrm{lb}}\}, \quad i \in I.$$

It holds that $S_{i,\mathrm{ub}} \geq S_i^*$ for all $i \in I$.

In the enumeration procedure, one searches for the best solution among all solutions $\mathbf{S}$ with $S_{i,\mathrm{lb}} \leq S_i \leq S_{i,\mathrm{ub}}$ for all $i \in I$.

Note that an alternative feasible solution (e.g., a solution generated by one of the two heuristics as described below), may lead to a tighter bound and thus to a faster enumeration algorithm.

Example 3.1 (continued). Via the above enumeration algorithm we can solve Problem (P) for our example. We find $(S_{1,\text{lb}}, S_{2,\text{lb}}, S_{3,\text{lb}}) = (2,1,4)$ and $C_{\text{P}}^{\text{lb}} = 1{,}541$ Euros per month as lower bounds for the optimal basestock levels and optimal costs. As upper bounds, we find $(S_{1,\text{ub}}, S_{2,\text{ub}}, S_{3,\text{ub}}) = (7,4,10)$ and $C_{\text{P}}^{\text{ub}} = 2{,}042$ Euros per month. Next, we obtain the optimal solution $\mathbf{S}^* = (3,2,5)$ with optimal costs $C_{\text{P}} = 1{,}750$ Euros per month and aggregate mean waiting times $\hat{W}_1(\mathbf{S}^*) = 0.118$ days and $\hat{W}_2(\mathbf{S}^*) = 0.148$ days. Under this optimal solution, $\hat{W}_1(\mathbf{S}^*)$ is much lower than the target $\hat{W}_1(\text{obj}) = 0.2$. This is an effect of the integrality of the solution space (a lower value for one of the base stock levels would lead to a violation of one of the aggregate mean waiting time constraints).

3.3.2 Greedy Heuristic

In Problem (P), we have an aggregate mean waiting time constraint per machine type $n \in N$. Generally, we have multiple machine types and thus multiple constraints. As in Chap. 2, generating efficient solutions for the corresponding multi-objective problem of Problem (P) would still be possible, but one would need large computation times and one gets many efficient solutions in an $(|N|+1)$-dimensional space. Therefore, we now follow a different path to obtain a good, feasible solution. We formulate again a greedy algorithm, but this time the logic behind the algorithm is based on Problem (P) itself.

First of all, we divide the set $\mathscr{S}$ of all solutions into a subset

$$\mathscr{S}^{\text{feas}} := \{\mathbf{S} \in \mathscr{S} | \hat{W}_n(\mathbf{S}) \leq \hat{W}_n^{\text{obj}} \text{ for all } n \in N\}$$

of feasible solutions and the subset $\mathscr{S} \setminus \mathscr{S}^{\text{feas}}$ of non-feasible solutions. Next, for each solution $\mathbf{S} \in \mathscr{S}$, we define

$$d(\mathbf{S}) := \sum_{n \in N} \left(\hat{W}_n(\mathbf{S}) - \hat{W}_n^{\text{obj}} \right)^+,$$

with $x^+ := \max\{0, x\}$ for all $x \in \mathbb{R}$, as the distance to the set $\mathscr{S}^{\text{feas}}$ of feasible solutions.

Now, a good, feasible solution for Problem (P) is obtained as follows. We first determine the solution $\mathbf{S}_{\text{lb}} = (S_{1,\text{lb}}, \ldots, S_{|I|,\text{lb}})$ (cf. (3.4)), which minimizes $C(\mathbf{S})$ over $\mathbf{S} \in \mathscr{S}$. If the target levels $\hat{W}_n^{\text{obj}}$ are relatively large, then $\mathbf{S}_{\text{lb}}$ may be a feasible solution, i.e., $\mathbf{S}_{\text{lb}} \in \mathscr{S}^{\text{feas}}$, and we can stop our algorithm because we have found an optimal solution. Generally $\mathbf{S}_{\text{lb}}$ will not be feasible, i.e., $\mathbf{S}_{\text{lb}} \in \mathscr{S} \setminus \mathscr{S}^{\text{feas}}$ and $d(\mathbf{S}_{\text{lb}}) > 0$, and then we move to the set of feasible solutions via greedy steps.

In each greedy step, we increase one of the basestock levels S_i of the current solution $\mathbf{S} = (S_1, \ldots, S_{|I|})$ (in the first step, $\mathbf{S} = \mathbf{S}_{\text{lb}}$). For each $i \in I$, we consider the increase in total costs and the decrease in the distance $d(\mathbf{S})$ to the set of feasible solutions if S_i would be increased by 1 unit. The increase in costs is equal to

$$\Delta C_i(S_i) = C_i(S_i + 1) - C_i(S_i) \tag{3.6}$$

(because we start with the solution $\mathbf{S}_{\text{lb}}$, $\Delta C_i(S_i) > 0$ in each greedy step). The decrease in the distance to the feasible solutions is given by $-\Delta_i d(\mathbf{S})$, where ($\mathbf{e}_i$ is the i-th unit vector with dimension $|I|$)

$$\begin{aligned} \Delta_i d(\mathbf{S}) &= d(\mathbf{S} + \mathbf{e}_i) - d(\mathbf{S}) \\ &= \sum_{n \in N} \left[\left(\hat{W}_n(\mathbf{S} + \mathbf{e}_i) - \hat{W}_n^{\text{obj}} \right)^+ - \left(\hat{W}_n(\mathbf{S}) - \hat{W}_n^{\text{obj}} \right)^+ \right]. \end{aligned} \tag{3.7}$$

The factor

$$\Gamma_i = \frac{-\Delta_i d(\mathbf{S})}{\Delta C_i(S_i)}$$

denotes the decrease in distance to the feasible solutions per unit of increase in costs. For the SKU i with the highest ratio Γ_i, we have the "biggest bang for the buck" and we increase the basestock level to $S_i + 1$.

We continue with the above greedy steps until we arrive at a feasible solution. We denote the costs of the solution generated by the greedy heuristic by C_{greedy}. The formal procedure is described in Algorithm 3.1.

Algorithm 3.1 *(Greedy heuristic)*

Step 1 Set $S_i := S_{i,\text{lb}}$ for all $i \in I$ (cf. (3.4));
  Determine $d(\mathbf{S})$
Step 2 While $d(\mathbf{S}) > 0$ do:
  a. Determine $\Delta C_i(S_i)$ for all $i \in I$ (cf. (3.6));
  b. Determine $\Delta_i d(\mathbf{S})$ for all $i \in I$ (cf. (3.7));
  c. Determine $\Gamma_i = -\Delta_i d(\mathbf{S}) / \Delta C_i(S_i)$ for all $i \in I$;
  d. $k := \arg\max\{\Gamma_i : i \in I\}$;
  e. $\mathbf{S} := \mathbf{S} + \mathbf{e}_k$

Example 3.1 (continued). By applying the greedy heuristic, we obtain the solution $\mathbf{S} = (3, 2, 5)$, which is equal to the optimal solution as generated via the enumeration algorithm. In Table 3.1, we have listed the solutions that are generated in Step 2 of the greedy heuristic.

Example 3.2. For this example, we assume that everything is the same as for Example 3.1, except the target levels for the aggregate mean waiting times. We now assume that $\hat{W}_1^{\text{obj}} = \hat{W}_2^{\text{obj}} = 0.1$ days per demand. Then, via enumeration, we obtain

Table 3.1 Solutions generated by the greedy heuristic for Example 3.1

Iteration	Γ_1	Γ_2	Γ_3	k	S_1	S_2	S_3	$\hat{W}_1(\mathbf{S})$	$\hat{W}_2(\mathbf{S})$	$C(\mathbf{S})$ (Euros)
0	–	–	–	–	2	1	4	0.319	0.477	0
1	$1.4 \cdot 10^{-2}$	$1.9 \cdot 10^{-3}$	$1.7 \cdot 10^{-3}$	1	3	1	4	0.157	0.477	1,551
2	0	$1.9 \cdot 10^{-3}$	$8.9 \cdot 10^{-4}$	2	3	2	4	0.157	0.191	1,700
3	0	$1.7 \cdot 10^{-4}$	$8.4 \cdot 10^{-4}$	3	3	2	5	0.118	0.148	1,750

Table 3.2 Solutions generated by the greedy heuristic for Example 3.2

Iteration	Γ_1	Γ_2	Γ_3	k	S_1	S_2	S_3	$\hat{W}_1(\mathbf{S})$	$\hat{W}_2(\mathbf{S})$	$C(\mathbf{S})$ (Euros)
0	–	–	–	–	2	1	4	0.329	0.477	1,542
1	$1.9 \cdot 10^{-2}$	$1.9 \cdot 10^{-3}$	$1.7 \cdot 10^{-3}$	1	3	1	4	0.157	0.477	1,551
2	$6.2 \cdot 10^{-4}$	$1.9 \cdot 10^{-3}$	$1.7 \cdot 10^{-3}$	2	3	2	4	0.157	0.191	1,700
3	$6.2 \cdot 10^{-4}$	$3.7 \cdot 10^{-4}$	$1.7 \cdot 10^{-3}$	3	3	2	5	0.118	0.148	1,750
4	$1.9 \cdot 10^{-4}$	$1.9 \cdot 10^{-4}$	$3.5 \cdot 10^{-4}$	3	3	2	6	0.104	0.132	1,835
5	$3.8 \cdot 10^{-5}$	$1.3 \cdot 10^{-4}$	$8.2 \cdot 10^{-5}$	2	3	3	6	0.104	0.035	2,084
6	$3.8 \cdot 10^{-5}$	0	$3.5 \cdot 10^{-5}$	1	4	3	6	0.034	0.035	2,176

the optimal solution $\mathbf{S}^* = (4,3,4)$ with optimal costs $C_P = 2{,}042$ Euros per month and aggregate mean waiting times $\hat{W}_1(\mathbf{S}) = 0.088$ days and $\hat{W}_2(\mathbf{S}) = 0.094$ days. Via the greedy heuristic, we obtain the solution $\mathbf{S} = (4,3,6)$ with costs $C(\mathbf{S}) = 2{,}176$ Euros per month and aggregate mean waiting times $\hat{W}_1(\mathbf{S}) = 0.034$ days and $\hat{W}_2(\mathbf{S}) = 0.035$ days; see Table 3.2. The costs of this solution are 6.6 % higher than the optimal costs. The greedy heuristic generates the solution $\mathbf{S} = (3,2,4)$ after two iterations, and then increases the base stock level S_3 of the relatively cheap and common SKU 3 to 6, after which it increases S_2 and S_1, respectively. Because it is most attractive to increase S_3 in the iterations 3 and 4, in the end one obtains a solution under which the aggregate mean waiting times are significantly smaller than their target levels. In this example with 3 SKU's, this leads to an optimality gap of 6.6 %. When the greedy heuristic is applied to problems with many SKU's, the amounts with which targets are exceeded by the last iteration steps are much smaller and one gets smaller optimality gaps; see also Sect. 3.4.

3.3.3 Dantzig-Wolfe Decomposition

In this section, we describe the application of Dantzig-Wolfe decomposition [3] to Problem (P), which leads to a lower bound for the optimal costs C_P and another heuristic solution.

3.3.3.1 Lower Bound

We first introduce a *Master Problem*, in which the variables of our original problem are expressed as convex combinations of columns that contain all possible values for the decision variables in the original problem. Let $K := \mathbb{N}_0$ denote the set of base stock policies for each of the SKU's $i \in I$. Let S_i^k, $i \in I, k \in K$, denote the base stock level of policy k for SKU i, and let $x_i^k \in \{0,1\}$, $i \in I, k \in K$, be a variable indicating whether policy k for SKU i is chosen ($x_i^k = 1$) or not ($x_i^k = 0$). Relaxing the integrality constraint on x_i^k, $i \in I, k \in K$, a suitable Master Problem related to Problem (P) is defined as follows:

$$
\begin{array}{llll}
\text{(MP)} & \min & \sum_{i \in I} \sum_{k \in K} C_i(S_i^k) x_i^k & \\
 & \text{subject to} & \sum_{i \in I} \sum_{k \in K} \frac{m_{i,n}}{M_n} W_i(S_i^k) x_i^k \leq \hat{W}_n^{\text{obj}}, \quad n \in N, & \text{(MP.1)} \\
 & & \sum_{k \in K} x_i^k = 1, \qquad i \in I, & \text{(MP.2)} \\
 & & x_i^k \geq 0, \qquad i \in I, k \in K. &
\end{array}
$$

The optimal costs of Problem (MP) are denoted by C_{MP}. Notice that the relaxation of the integrality condition on x_i^k, $i \in I, k \in K$, in Problem (MP) allows for fractional values of x_i^k, $i \in I, k \in K$, and thus corresponds to allowing *randomized policies*. Therefore, C_{MP} constitutes a lower bound on C_{P}.

Besides Problem (MP), a Restricted Master Problem, Problem (RMP), is defined that for each SKU $i \in I$ only considers a small subset $K_i \subseteq K$ of columns (policies). The optimal cost of Problem (RMP) is denoted by C_{RMP}. For each SKU $i \in I$, let K_i initially consist of one policy k, with $S_i^k := \min\{x \in \mathbb{N}_0 | W_i(x) \leq \hat{W}_{\min}^{\text{obj}}\}$ (similar to (3.5)). This choice of initial policies constitutes a feasible solution for Problem (RMP). That S_i^k is finite for each $i \in I$, can be seen as follows. The Erlang loss probability, $1 - \beta_i(S_i)$, is strictly decreasing and strictly convex on its entire domain $\mathbb{N}_0$ (see Karush [5]; notice that we used this property also in Sect. 2.9), and, obviously, $\beta_i(\infty) = 1$. Thus, $W_i(S_i)$ is strictly decreasing and strictly convex on its entire domain $\mathbb{N}_0$, and $W_i(\infty) = 0$. So, there is a finite x for which $W_i(x) \leq \hat{W}_{\min}^{\text{obj}}$, and thus S_i^k is finite.

After solving Problem (RMP) with the $|I|$ initial policies, we are interested in policies that have not yet been considered, but that would improve the solution of Problem (RMP) if they were added. To check whether such policies exist, we solve, for each SKU $i \in I$, a so-called *column generation subproblem* that for that SKU generates a policy with the lowest reduced cost coefficient. Given an optimal solution for Problem (RMP), let $u_n \leq 0$, $n \in N$, denote the dual variables (these are directly related to the shadow prices: 'dual variable' $=-$ 'shadow price') related to the $|J|$ service level constraints (MP.1), and let v_i, $i \in I$, denote the dual variables related to the $|I|$ constraints (MP.2). Then, for our Problem (RMP), the column generation subproblem for an SKU $i \in I$, is as follows:

$$\text{(SUB}(i))\ \min \qquad C_i(S_i) - \sum_{n \in N} u_n \frac{m_{i,n}}{M_n} W_i(S_i) - v_i$$

$$\text{subject to } S_i \in \mathbb{N}_0.$$

Let $S_{\mathrm{SUB}(i)}$ denote an optimal policy (base stock level) for Problem (SUB(i)), and let $C_{\mathrm{SUB}(i)}$ denote the costs of an optimal solution of Problem (SUB(i)), i.e., the lowest reduced cost coefficient, which also can be interpreted as the degree of violation of the corresponding constraint in the dual problem of Problem (MP). Recall that $\beta_i(S_i)$ is strictly increasing and strictly concave on its entire domain $\mathbb{N}_0$, since the Erlang loss probability is strictly decreasing and strictly convex. The function $C_i(S_i)$ consists of terms that are linear in S_i and terms having $\beta_i(S_i)$ with a non-positive coefficient (see Eq. (3.3)). The objective function of Problem (SUB(i)) further consists of another term containing $\beta_i(S_i)$ (via $W_i(S_i)$; see Eq. (3.2)) with a non-positive coefficient u_n, and a constant term v_i. Thus, the objective function of Problem (SUB(i)) is convex in S_i. This implies that Problem (SUB(i)) can be solved in a straightforward way. If there exists an optimal policy $S_{\mathrm{SUB}(i)}$ with a negative reduced cost coefficient $C_{\mathrm{SUB}(i)}$ for SKU i, this policy is added to K_i.

As long as a policy with a negative reduced cost coefficient exists for one or more SKU's, adding columns to Problem (RMP) and solving Problem (RMP) is done iteratively. If for none of the SKU's $i \in I$ a policy with negative reduced cost can be found, the obtained solution for Problem (RMP) is also optimal for Problem (MP).

Regarding the convergence of this iterative procedure, we can say the following. With our choice to let each K_i, $i \in I$, initially consist of one policy k with $S_i^k := \min\{x \in \mathbb{N}_0 | W_i(x) \leq \hat{W}_{\min}^{\mathrm{obj}}\}$, we cannot formally prove that convergence is obtained. Under a start with $S_i^k := \min\{x \in \mathbb{N}_0 | W_i(x) \leq \frac{1}{2}\hat{W}_{\min}^{\mathrm{obj}}\}$, we would have an initial solution for Problem (MP) that satisfies the constraints (MP.1) in the strict ($<$) sense for each $n \in N$. That solution is a so-called nondegenerate basic feasible solution for Problem (MP) and under such a starting solution one can guarantee that the column generation algorithm converges; see [4]. It is common however that a column generation algorithm is started with a solution that satisfies all constraints in the non-strict sense, and then generally the algorithm also converges. With our choice for the starting solution, we are in the latter situation. We applied our procedure to many problem instances, and found that the algorithm always converged.

In total, $|I| + |N|$ variables are in the basis. Notice that at most $|N|$ SKU's will have fractional x_i^k-values because the constraints (MP.2) require that for each i at least one x_i^k is a basic variable. Furthermore, notice that the number of service level constraints (MP.1) that are satisfied with equality is at least equal to the number of SKU's that have fractional x_i^k-values, again because of the number of variables in the basis.

3.3.3.2 Heuristic Solution

After a lower bound has been found for Problem (P), as described above, a good feasible solution for Problem (P) can be determined as follows.

If none of the x_i^k-values in the solution of Problem (MP) is fractional, the obtained solution of Problem (MP) is feasible for Problem (P) as well, and it is optimal.

If fractional x_i^k-values do occur, however, we need to apply some further steps. Since the number of fractional x_i^k-values is at most $|N|$, and, usually, $|N|$ is very small compared to $|I|$, it would be reasonable to select for each of these few SKU's, the policy with a non-zero x_i^k-value and the highest base stock level. Obviously, this results in a feasible solution for Problem (P), but it may lead to a poor feasible solution (i.e., a solution that is far from optimal) if one of these SKU's happens to be expensive. Therefore, another procedure is proposed to obtain an upper bound.

Select for each SKU $i \in I$ the policy with a non-zero x_i^k-value and the lowest base stock level S_i^k, and refer to this policy as policy k'. Notice that $C_i(D) \geq C_i(S_i^{k'})$ for $D > S_i^{k'}, D \in \mathbb{N}$. This holds because $\beta_i(D) \geq \beta_i(S_i^{k'})$, and thus if $C_i(D) < C_i(S_i^{k'})$, a feasible solution would exist with $x_i^{k'} = 0$ that has lower costs than the current solution of Problem (MP). This would contradict the choice of k'. Let $S_i := S_i^{k'}$ and $\mathbf{S} := (S_1, \ldots, S_{|I|})$.

Notice that $\hat{W}_n(\mathbf{S}) > \hat{W}_n^{\text{obj}}$ for at least one n, i.e., that policy $\mathbf{S}$ is infeasible. This holds because $x_i^{k'}$ is fractional for at least one i. It can be verified that for this i it holds that $\beta_i(S_i^{k'}) < \beta_i(S_i^l)$, and thus that $W_i(S_i^{k'}) > W_i(S_i^l)$, and furthermore that $C_i(S_i^{k'}) < C_i(S_i^l)$, with l denoting any other policy with fractional x_i^l-value, and that $x_i^{k'} = 1$ would lead to an infeasible solution, i.e., a solution with $\hat{W}_n(\mathbf{S}) > \hat{W}_n^{\text{obj}}$ for at least one n.

Starting with the policy $\mathbf{S}$, we generate a feasible solution by greedy steps, as in the greedy heuristic; i.e., we take $\mathbf{S}$ as a starting solution and apply Step 2 of Algorithm 3.1. Generally, this will give a close-to-optimal solution. We denote the costs of the solution generated by the Dantzig-Wolfe heuristic by C_{DW}. For both the greedy heuristic and the Dantzig-Wolfe heuristic, we measure the quality by the gap

$$G_k := \frac{C_k - C_{\text{MP}}}{C_{\text{MP}}}, \quad k \in \{\text{greedy}, \text{DW}\},$$

which measures the distance to the lower bound and thus constitutes an upper bound for the optimality gap. This gap is an important indicator of the quality of heuristics. We use G_{greedy} and G_{DW} to evaluate the performance of the heuristics in Sect. 3.4.

Example 3.1 (continued). By applying the Dantzig-Wolfe decomposition as described above, we obtain $C_{\text{MP}} = 1{,}743$ Euros per month as a lower bound for the optimal costs C_{P}, and $\mathbf{S} = (3,2,5)$ as heuristic solution with total average costs $C_{\text{DW}} = 1{,}750$. The latter solution is equal to the solution obtained by the greedy algorithm in Sect. 3.3.2, and also equal to the optimal solution for this example. The relative gap between the lower bound and the costs of solution $\mathbf{S} = (3,2,5)$ is 0.4 %, which is quite small.

Example 3.2 (continued). By applying the Dantzig-Wolfe decomposition as described above, we obtain $C_{\mathrm{MP}} = 1{,}895$ Euros per month as a lower bound for the optimal costs C_{P}, and $\mathbf{S} = (4,3,6)$ as heuristic solution with total average costs $C_{\mathrm{DW}} = 2{,}176$. The latter solution is equal to the solution obtained by the greedy algorithm in Sect. 3.3.3. While the optimality gap of $\mathbf{S} = (4,3,6)$ is 6.6 % (see Example 3.2 in Sect. 3.3.2), the gap with the lower bound, denoted by $G_{\mathrm{DW}} = G_{\mathrm{greedy}}$ is now 14.8 %. This is a large gap which one sometimes obtains for small instances. (We found that a gap of this order of magnitude can also be obtained for some large, symmetric instances, in which case the integrality character of the problem stays strong; in practice, such symmetric instances will not occur.)

3.4 Computational Results

In this section, we execute two computational experiments. In the first experiment, we investigate the quality of the solutions generated by the greedy heuristic and the Dantzig-Wolfe heuristic. The setup of this experiment is described in Sect. 3.4.1 and the results are discussed in Sect. 3.4.2. In the second experiment, presented in Sect. 3.4.3, we investigate the benefits of commonality.

3.4.1 Setup for Experiment 1

One of the factors that we want to vary in the first experiment is the so-called commonality percentage CP_n per machine type n. Define the set I_n as the set of active SKU's for machine group n. Active SKU's are SKU's that occur in the configuration of a machine type and have positive demand rates; i.e.,

$$I_n = \{i \in I | m_{i,n} > 0\}, \quad n \in N.$$

The commonality percentage CP_n for machine type n is defined as the fraction of the SKU's in I_n that also receive demand from at least one other machine type; i.e.,

$$CP_n := \frac{|\{i \in I_n | m_{i,n} < \mu_i\}|}{|I_n|}, \quad n \in N.$$

By an appropriate construction of the demand rates $m_{i,n}$, we can control the values of the number of elements in the sets I_n and the commonality percentages CP_n. We construct our problem instances such that $|I_n| = |I_1|$ and $CP_n = CP_1$ for all $n \in N$.

In the first experiment, we consider the values 20 and 100 for the sets $|I_n|$, the numbers 0.2, 0.5, and 0.8 for the commonality percentages CP_n and the values 2 and 5 for the number of machine types $|N|$. For the targets $\hat{W}_n^{\mathrm{obj}}$ for the aggregate mean waiting times, we consider 5 settings, which we denote by (i)–(v). For the instances with $|N| = 2$ machine types, the settings for $(\hat{W}_1^{\mathrm{obj}}, \hat{W}_2^{\mathrm{obj}})$ are:

$$\begin{aligned}
\text{(i):}&\quad (0.025, 0.025),\\
\text{(ii):}&\quad (0.025, 0.050),\\
\text{(iii):}&\quad (0.050, 0.050),\\
\text{(iv):}&\quad (0.050, 0.100),\\
\text{(v):}&\quad (0.100, 0.100).
\end{aligned}$$

For the instances with $|N| = 5$ machine types, the settings for $(\hat{W}_1^{\text{obj}}, \ldots, \hat{W}_5^{\text{obj}})$ are:

$$\begin{aligned}
\text{(i):}&\quad (0.025, 0.025, 0.025, 0.025, 0.025),\\
\text{(ii):}&\quad (0.025, 0.025, 0.0375, 0.050, 0.050),\\
\text{(iii):}&\quad (0.050, 0.050, 0.050, 0.050, 0.050),\\
\text{(iv):}&\quad (0.050, 0.050, 0.075, 0.100, 0.100),\\
\text{(v):}&\quad (0.100, 0.100, 0.100, 0.100, 0.100).
\end{aligned}$$

The demand rates $m_{i,n}$ are randomly generated as follows. For each setting of the parameters explained above, we generate 10 random problem instances. For each problem instance, we first define the common SKU's. The number of common SKU's is $|I_n|CP_n$, which is always an integer value under the above choices for $|I_n|$ and CP_n (e.g., if $|I_n| = 20$ and $CP_n = 0.2$, then we have $|I_n|CP_n = 4$ common SKU's). The number of specific SKU's for a machine group is $|I_n|(1 - CP_n)$, and the total number of SKU's is

$$|I| = |I_n|CP_n + |N||I_n|(1 - CP_n)$$

(e.g., if $|I_n| = 20$, $|N| = 2$ and $CP_n = 0.2$, then $|I| = 36$). We define the SKU's $1, \ldots, |I_n|CP_n$ as the common SKU's. The SKU's $|I_n|CP_n + 1, \ldots, |I_n|$ are the specific SKU's for machine type 1, the SKU's $|I_n| + 1, \ldots, |I_n| + |I_n|(1 - CP_n)$ are the specific SKU's for machine group 2, and so on. Next, for each specific SKU, demand rate $m_{i,n}$ is generated from a uniform distribution $U[0.005, 0.100]$. For each common SKU, we first generate a random demand rate from a uniform distribution $U[0.005, 0.100]$. Then, we multiply each of these random demand rates with a factor generated from a second uniform distribution $U[0.5, 1.5]$ to obtain demand rate $m_{i,n}$. In this way, we keep a positive correlation between the demand rates from the various machine types but we avoid that the demand rates are identical. This reflects what one generally also has in practice.

The inventory holding cost parameters c_i^{h} (in Euros per day per part) are generated from a uniform distribution $U[0.1, 10]$. Here, each value is drawn independently for all SKU's $i \in I$.

Further, we take equal mean regular replenishment leadtimes t_i, equal mean emergency shipment times t_i^{em}, and equal emergency shipment costs c_i^{em} for all $i \in I$. For the t_i, we take 20 days. For the t_i^{em}, we take 1 day. We set c_i^{em} equal to 750 Euros. In total, we have $10 \times 2 \times 2 \times 3 \times 5 = 600$ problem instances. An overview of the test bed is given in Table 3.3.

For each instance, we apply the greedy heuristic and Dantzig-Wolfe decomposition, which give us the lower bound C_{MP} for the optimal costs, the costs C_{greedy}

Table 3.3 Parameter settings for Experiment 1

Parameter	Number of choices	Values
Number of active SKU's per machine type ($\|I_n\|$)	2	20, 100
Number of machine types ($\|N\|$)	2	2, 5
Commonality percentage (CP_n)	3	0.2, 0.5, 0.8
Target waiting times ($\hat{W}_n^{\text{obj}}$; in days)	5	(i), (ii), (iii), (iv), (v)
Emergency shipm. costs (c_i^{em}; in Euros)	1	750
Mean emergency shipm. time (t_i^{em}; in days)	1	1
Mean repl. lead time (t_i; in days)	1	20
Demand rate ($m_{i,n}$; in units per day)	1	$U[0.005, 0.100](\times U[0.5, 1.5])$
Inv. holding costs (c_i^{h}; in Euros per unit per day)	1	$U[0.1, 10]$

of the solution generated by the greedy heuristic, and the costs C_{DW} of the solution generated by the Dantzig-Wolfe heuristic. The Dantzig-Wolfe decomposition has been implemented in AIMMS 3.4, and CPLEX 12.6 is used as solver for the linear programming problems. The greedy heuristic has been implemented in Delphi XE7. The algorithms are run on a 2.60 GHz Intel Core i5 3320M notebook with 4 GB RAM.

3.4.2 Quality of the Heuristics

The results of the experiment are listed in Table 3.4. In this table, we present the performance of each heuristic method in terms of the average and maximum value of the relative gaps G_{greedy} and G_{DW}. We also investigate the ratio of instances in which the greedy heuristic stops after Step 1, denoted by R^0_{greedy}. In these instances, the solution $S_{\text{lb}} = (S_{1,\text{lb}}, \ldots, S_{|I|,\text{lb}})$ which minimizes $C(\mathbf{S})$, satisfies already the aggregate mean waiting time constraints for all machine types, i.e., we can say that these constraints are not binding. In Table 3.4, we distinguish subsets of instances with the same value for a specific input parameter and in the bottom line the results for all instances together are presented. The performance of each heuristic in terms of the computation time, denoted by CPU_{greedy} and CPU_{DW}, respectively, is presented in Table 3.5.

The main observations drawn from the experiment are as follows:

- Both the Dantzig-Wolfe heuristic and the greedy heuristic perform very well. The averages of G_{greedy} and G_{DW} over all instances are around 0.1 % and the maximum value is 1.225 %.
- The average and maximum G_{greedy} and G_{DW} decrease strongly when the number of active SKU's per machine type increases (then $|I|$ increases as well). This is due to a decreasing effect of the integrality character of Problem (P) for increasing numbers of SKU's. This is an important observation because the number of SKU's usually is large in real-life situations.

Table 3.4 Results for Experiment 1

Parameter		G_{DW} (%)		G_{greedy} (%)		R^0_{greedy} (%)
		Avg	Max	Avg	Max	Avg
$\|I_n\|$	20	0.175	1.225	0.190	1.225	23.0
	100	0.015	0.068	0.022	0.197	27.3
$\|N\|$	2	0.116	1.225	0.132	1.225	21.3
	5	0.073	0.561	0.080	0.635	29.0
CP_n	0.2	0.120	1.225	0.123	1.225	18.0
	0.5	0.090	0.614	0.107	0.844	20.0
	0.8	0.074	0.848	0.088	0.848	37.5
$\hat{W}_n^{\text{obj}}$	(i)	0.177	1.225	0.194	1.225	00.0
	(ii)	0.156	1.225	0.186	1.225	00.0
	(iii)	0.078	0.451	0.082	0.604	14.2
	(iv)	0.062	0.434	0.067	0.604	15.0
	(v)	0.001	0.062	0.001	0.062	96.7
All instances		0.095	1.225	0.106	1.225	25.2

Table 3.5 Computation times for Experiment 1 (in seconds)

Parameter		CPU_{DW}		CPU_{greedy}	
		Avg	Max	Avg	Max
$\|I_n\|$	20	0.143	0.430	0.011	0.099
	100	0.665	2.100	0.203	0.998
$\|N\|$	2	0.226	0.660	0.062	0.394
	5	0.582	2.100	0.152	0.998
CP_n	0.2	0.479	2.100	0.126	0.995
	0.5	0.355	1.690	0.108	0.998
	0.8	0.377	1.570	0.088	0.935
$\hat{W}_n^{\text{obj}}$	(i)	0.557	2.100	0.210	0.964
	(ii)	0.470	1.950	0.170	0.998
	(iii)	0.345	1.280	0.074	0.995
	(iv)	0.377	1.230	0.080	0.919
	(v)	0.271	0.760	0.001	0.002
All instances		0.404	2.100	0.107	0.998

- The average G_{greedy} and G_{DW} also decrease strongly when the targets for the aggregate mean waiting times increase. The explanation is that then the tightness of the aggregate mean waiting time constraints decreases and they become non-binding for more machine types. For the greedy algorithm, this leads to more instances in which the greedy algorithm stops after Step 1. This is confirmed by the increasing values for R^0_{greedy}. In these instances one obtains a zero gap. In these instances, the Dantzig-Wolfe heuristic generally also finds the optimal solution, and thus a similar effect is obtained for G_{DW}.
- Further, the average G_{greedy} and G_{DW} are decreasing when the number of machine types $|N|$ or the commonality percentage CP_n increases. An increasing number of machine types leads to higher demand rates for the common SKU's, and that seems to give a less strong effect of the integrality character of Problem (P). A higher commonality percentage implies that one has more common SKU's and that also gives a less strong effect of the integrality character.

- From Table 3.5, we can conclude that both heuristics are very efficient in terms of computation time. For almost all instances, CPU times are less than a second. As expected, the computation time for the Dantzig-Wolfe heuristic is slightly higher than for the greedy heuristic. The computation times for both heuristics increase significantly when the number of active SKU's per machine type or the number of machines increases. Further, the computation time for the greedy algorithm decreases for increasing values of CP_n and $\hat{W}_n^{\text{obj}}$. For the Dantzig-Wolfe heuristic, the computation time also decreases for increasing values of $\hat{W}_n^{\text{obj}}$, but there is no monotonic behavior as a function of CP_n.

The general conclusion is that both the greedy heuristic and the Dantzig-Wolfe heuristic is effective and efficient.

3.4.3 Benefits of Commonality

In our second experiment, we investigate the benefits of commonality. We arbitrarily chose 100 SKU's from one machine type in a data set that we obtained from ASML. We checked their representativeness by plotting failure rates versus prices. This showed a pattern similar to the complete set of SKU's for that machine type. For these 100 SKU's, we studied several scenarios for situations with 2 and 5 machine types. We let all machine types have these 100 SKU's, with their failure rates and prices (regarding the failure rates for a SKU i, each $m_{i,n}$ is taken equal to the corresponding rate in the data set). In these scenarios, we varied the CP_n-value from 0 % to 100 % in steps of 20 % (within a scenario we assumed CP_n equal for all n). Furthermore, we varied the commonality setting, indicating whether commonality occurs in cheap or expensive SKU's or equally distributed over all SKU's. According to these settings, we declared an SKU in a scenario either as completely common (i.e., occurring in all machine types) or as specific per machine type. We set $t_i = 14$ days, $t_i^{\text{em}} = 1$ day, and $c_i^{\text{em}} = 750$ Euro. The inventory holding cost c_i^{h} for a SKU i is set at 25 % of its price on a yearly basis, and dividing by 365 gives the inventory holding cost in Euros per unit per day. We set $\hat{W}_n^{\text{obj}} = 0.05$ for all n. In total, we have $2 \times 6 \times 3 = 36$ instances in our test bed.

For each instance, we compute the costs when a shared stock is used for all machine types together and the costs when a separate stock is kept for each machine type. Due to how our instances were defined, the costs of the latter case are equal to the costs for the equivalent case with 0 % commonality. For both the shared and separate stock case, we use the solution obtained by the Dantzig-Wolfe heuristic. In Figs. 3.1 and 3.2, we show the total costs under shared stocks relative to the total costs under separate stocks (i.e., 0 % commonality). Notice that at 100 % commonality, there are no differences between the three commonality settings and thus the same costs are obtained.

From Figs. 3.1 and 3.2, it can be seen that enormous savings can be obtained if the commonality percentage CP_n is high. Furthermore, if the number of machine

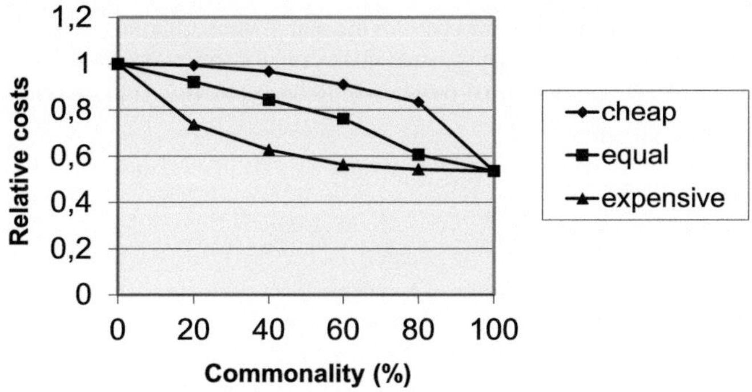

Fig. 3.1 Relative costs under shared stocks for two machine types

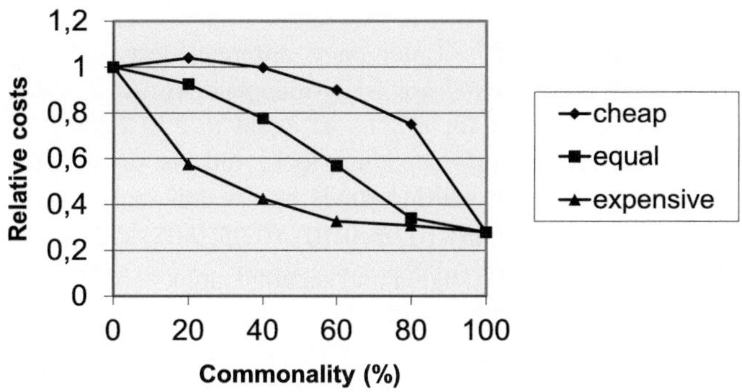

Fig. 3.2 Relative costs under shared stocks for five machine types

types increases, the benefits of using shared instead of separate stocks increases as well. Third, commonality in expensive SKU's leads to a much larger reduction in costs than commonality in cheap SKU's. Even if CP_n is about 40 %, only a small saving can be obtained if the commonality occurs in cheap SKU's only.

3.5 Case Study: ASML

In the case study, we consider 8 data sets of ASML, corresponding to 8 different local warehouses. For each data set, we have $|N| = 2$ machine types. The average number of SKU's per machine type is about 700, and varies between 400 and 1,000. Demand rates are low, on average 0.25 per year, and vary between 0.0005 and 40 per year. The relative size of the machine types, expressed in terms of M_1/M_2, and CP_1 and CP_2 are given in Table 3.6. On average, CP_n is 0.19. For the 10 % most expensive SKU's per machine type, the commonality percentages are about the same as the depicted values, which gives us an indication that for all machine types, the occurrence of commonality is equally

Table 3.6 Results case study: spare parts provisioning costs for the shared stock situation, depicted as fraction of the spare parts provisioning costs in the separate stock case, for various target waiting time settings $(\hat{W}_1^{\text{obj}}, \hat{W}_2^{\text{obj}})$ ((i): (0.10,0.10), (ii): (0.10,0.05), (iii): (0.05,0.10), (iv): (0.05,0.05))

Data set	M_1/M_2	CP_1	CP_2	Target waiting times (i)	(ii)	(iii)	(iv)
1	8.95	0.12	0.23	0.96	0.93	0.96	0.94
2	0.40	0.10	0.12	0.96	0.97	0.96	0.97
3	33.22	0.12	0.26	0.93	0.94	0.93	0.94
4	2.37	0.12	0.25	0.93	0.93	0.93	0.92
5	0.75	0.10	0.19	0.97	0.97	0.96	0.96
6	0.57	0.19	0.32	0.93	0.94	0.91	0.91
7	1.25	0.18	0.30	0.94	0.95	0.93	0.93
8	10.00	0.12	0.26	0.92	0.93	0.92	0.93

distributed over cheap and expensive SKU's. Emergency shipment costs c_i^{em} and times t_i^{em} and regular replenishment times t_i are SKU-independent. At ASML, $t_i = 14$ days, and $t_i^{\text{em}} = 1$ day. The value for c_i^{em} is set equal to 750 Euro. For each data set, we compare using shared and separate stock, and we do that for four settings of the target aggregate mean waiting times that match real-life targets: $(\hat{W}_1^{\text{obj}}, \hat{W}_2^{\text{obj}}) \in \{(0.10,0.10),(0.10,0.05),(0.05,0.10),(0.05,0.05)\}$. The targets $\hat{W}_n^{\text{obj}}$ are expressed in days. For both the shared and separate stock case, we use the solution obtained by the Dantzig-Wolfe heuristic.

In Table 3.6, the spare parts provisioning costs for the shared stock situation are depicted as fractions of the spare parts provisioning costs in the separate stock case, for all 8 data sets and the 4 different settings for the target waiting times. On average 6 % can be saved in spare parts provisioning costs in the data sets obtained from ASML (this average is computed over 32 combinations of the data sets and target waiting time). Target waiting times and the relative sizes of the machine types seem to have little influence on this.

On average, it took 13 s per instance to run the Dantzig-Wolfe decomposition on a Pentium 4 computer, and the maximum computation time was 26 s. We observed that the method generates quite good solutions: the gap G_{DW} is on average 0.06 % and at most 0.3 % in the considered instances.

3.6 Concluding Remarks

Commonality, i.e., the fact that parts are used in more than one machine type, has received attention in several papers, mainly in the context of assemble-to-order systems (see [7, 8], and the references therein). Both works distinguish two streams of research with respect to commonality. One stream of research is on the design of the system or product: common components can be beneficial in terms of risk sharing, but in general have higher cost and lower efficiency than dedicated components. Another stream of research covers efficient operation, given the design of the system. Multi-item spare parts models show some similarity to models for assemble-to-order systems. Both have, in the general form, multiple machines or products that consist of multiple components, that can be common, and both focus on service measures at the machine level. However, a main difference is that in an assemble-to-order model a requested product can be assembled if *all* items are available (coupled demands), while in multi-item spare parts models demands occur for individual items. With respect to commonality in multi-item models, we also would like to refer to [1, 2]. These papers discuss how commonality can be dealt with in two different periodic review models, but they do not study cost benefits of incorporation of commonality in the models.

In this chapter, we have studied the multi-item spare parts problem with commonality and we have assumed basestock control. We have derived two efficient and effective heuristics, a greedy heuristic and a Dantzig-Wolfe heuristic, and we showed the benefits of commonality (by comparing one shared stock to separate stocks per machine type). Further, we executed a case study at ASML. The results on the benefits of commonality (including the use of Dantzig-Wolfe decomposition) and the case study stem from [6].

Problems

3.1. Consider the same problem as in Example 3.1, but the price of SKU 2 is now equal to 8,000 Euros instead of 20,000 Euros. Apply the greedy heuristic. Which solution $\mathbf{S}$ do you obtain, and what are the corresponding values for the aggregate mean waiting times $\hat{W}_n(\mathbf{S})$, $n = 1,2$, and the average costs $C(\mathbf{S})$?

3.2. Consider the problem in Example 3.1. Instead of having *shared stock*, we may also have *separate stocks* for each machine type. In this problem, we compute the solution that one then obtains. Under separate stocks, one can apply the model and greedy algorithm of Sect. 2.9.

(a) We first look at solving the separate stock problem for machine type 1. The first step is to describe the problem instance for the model of Sect. 2.9. It is convenient to number the SKU's as SKU A and SKU B in this problem instance, where SKU A corresponds to SKU 1 and SKU B corresponds to SKU 3. Describe the problem instance.
(b) We follow the notation of Sect. 2.9. For $i = A,B$, list in a table the values of $\rho_i^{S_i}/(S_i!)$, $1-\beta_i(S_i)$, $W_i(S_i)$, and $\hat{C}_i(S_i)$ for $0 \le S_i \le 4$.
(c) Apply the greedy algorithm of Sect. 2.9. What solution do you obtain, and what are the corresponding values for the average costs and the aggregate mean waiting time?
(d) The steps in the parts (a)–(c) also have to be applied for machine type 2. This results in a basestock level of 3 parts for both SKU 2 and SKU 3. The corresponding average costs are equal to 1,245 Euros per month and the aggregate mean waiting time equals 0.057 days per month. What are the total average

costs under separate stocks, and how large is the increase in costs compared to the optimal solution under shared stock?

3.3. In the model of this chapter, it is assumed that an emergency shipment is used in case of a stockout. Let us now assume that we have backordering in case of a stockout, as in Chap. 2. Suppose that we still have an aggregate mean waiting time constraint per machine type, and that we want to minimize the total inventory holding costs. Formulate the minimization problem that we then obtain, and give formulas for all functions used in this formulation.

3.4. Consider the model of this chapter, but with an aggregate fill rate constraint instead of an aggregate mean waiting time constraint per machine type. In Problem (P), the constraints $\hat{W}_n(\mathbf{S}) \leq \hat{W}_n^{\text{obj}}$, $n \in N$, are then replaced by $\hat{\beta}_n(\mathbf{S}) \geq \hat{\beta}_n^{\text{obj}}$, $n \in N$, with

$$\hat{\beta}_n(\mathbf{S}) = \sum_{i \in I} \frac{m_{i,n}}{M_n} \beta_i(S_i), \quad n \in N.$$

Formulate an appropriate greedy heuristic for this adapted problem. I.e., follow the same logic as in Sect. 3.3.2 and denote what changes for this adapted problem.

Notice: In the special case that all average emergency shipment times t_i^{em} are equal, Problem (P) is equivalent to a problem with aggregate fill rate constraints; see Remark 3.1. Here, we are interested in the general case.

References

1. Cohen, M.A., Kleindorfer, P.R., Lee, H.L.: Near-optimal service constrained stocking policies for spare parts. Oper. Res. **37**, 104–117 (1989)
2. Cohen, M.A., Kleindorfer, P.R., Lee, H.L.: Multi-item service constrained (s, S) policies for spare parts logistics systems. Nav. Res. Logist. **39**, 561–577 (1992)
3. Dantzig, G.B., Wolfe, P.: Decomposition principle for linear programs. Oper. Res. **8**, 101–111 (1960)
4. Dantzig, G.B.: Linear Programming and Extensions. Princeton University Press, Princeton (1963)
5. Karush, W.: A queueing model for an inventory problem. Oper. Res. **5**, 693–703 (1957)
6. Kranenburg, A.A., Van Houtum, G.J.: Effect of commonality on spare parts provisioning costs for capital goods. Int. J. Prod. Econ. **108**, 221–227 (2007)[1]
7. Song, J.S., Zipkin, P.: Supply chain operations: assemble-to-order systems. In: De Kok, A.G., Graves, S.C. (eds.) Handbooks in Operations Research and Management Science, vol. 11, pp. 561–596. Elsevier, Amsterdam (2003)
8. Van Mieghem, J.A.: Note–commonality strategies: value drivers and equivalence with flexible capacity and inventory substitution. Manag. Sci. **50**, 419–423 (2004)

[1] This chapter has been reprinted from Kranenburg and Van Houtum [6] with kind permission from Elsevier.

Chapter 4
Service Differentiation

4.1 Introduction

In this chapter, we consider a single location model for the support of machines of the same type. The machines belong to groups with different targets for the aggregate mean waiting time. Machines in the highest group (lowest target level for the aggregate mean waiting time) belong to factories where the availability of the machines is most critical. For machines in the second group, the availability of the machines is somewhat less critical, and so on. The higher the machine group of a certain machine, the higher the price that is paid for the spare parts provisioning (either via a service contract or via internal costs within a company).

A common way to realize differentiated aggregate mean waiting times is via so-called *critical levels* (cf. Veinott [22]). A critical level is defined per SKU per machine group. Demands for a SKU by a given machine group are satisfied from stock as long as the on-hand stock is above the critical level, and otherwise the demand is satisfied by an emergency shipment. Hence, the higher the critical level the worse the service is. For the higher machine groups, we can define low critical levels (for the highest machine group all critical levels will be set equal to 0), and for lower machine groups, we can define high critical levels. So, via the critical levels we differentiate between demands from different machine groups. For the replenishments of the stock, we will assume basestock policies.

We will apply Dantzig-Wolfe decomposition, which will give us a lower bound for the optimal costs and a heuristic solution. Contrary to what we do in most other chapters in this book, in this chapter no greedy heuristic will be considered. Because of the presence of the critical levels, it is harder to define a good greedy heuristic. So far, such heuristics have not been studied in the literature.

Without using critical levels, all demands are treated in the same way. One then can meet all constraints by using the model of Chap. 3. If all machine groups have the same mix of demand rates (i.e., for each pair of machine groups, there is a SKU-independent constant such that the demand rates of one group are obtained

G.J. van Houtum, B. Kranenburg, *Spare Parts Inventory Control under System Availability Constraints*, International Series in Operations Research & Management Science 227, DOI 10.1007/978-1-4899-7609-3_4

by multiplying the demand rates of the other group by this constant), then using the model of Chap. 3 is equivalent to aggregating all machine groups into one group and using the target level of the highest machine group for this aggregated group. The policy that one then obtains is called a *round-up policy* (cf. [8]). In a computational experiment, we will compare the use of critical level policies to the use of round-up policies. We will also do this in a case study at ASML.

Although the focus of this chapter is on machine groups where all machines are of the same machine type, the model of this chapter can also be used when machines are of multiple machine types. One then gets a kind of mix of the situations of this and the previous chapter. One may distinguish multiple machine types and multiple target levels for the aggregate mean waiting time, and it seems most natural to define a machine group for each combination of a machine type and target level; see Remark 4.2.

This chapter is organized as follows. The model is described in Sect. 4.2. Next, before we apply Dantzig-Wolfe decomposition, we analyze an underlying single item problem in Sect. 4.3. The single item problem is the problem that has to be solved per SKU in order to generate new columns in the Dantzig-Wolfe method. Next, the Dantzig-Wolfe decomposition itself is described in Sect. 4.4, and the computational experiment is executed in Sect. 4.5. In Sect. 4.6, we apply our model to a case at ASML. Finally, the chapter is concluded in Sect. 4.7.

Significant parts of this chapter are based on [15].

4.2 Model

Consider machines of the same machine type that are installed at customers in one region. The machines consist of multiple critical components that are subject to failures. Each component, also referred to as stock-keeping unit (SKU), is assumed to be either a consumable or a repairable component. In the rest of the chapter, we use the terminology of repairables, but the whole approach works also for consumables. Spare parts for the SKU's are kept on stock in one warehouse.

The machines differ with respect to their criticality in the production processes of the factories where they are installed. Based on that criticality, the machines are divided into groups. Each group has its own target level for the aggregate mean waiting time.

Let I denote the set of SKU's, which are numbered $1,\dots,|I|$; we assume that $|I| \geq 1$. Let N denote the set of machine groups, with $|N| \geq 1$. The group with the lowest target for the aggregate mean waiting time is numbered as group 1, the group with the second lowest target is group 2, and so on. For each SKU $i \in I$ and machine group $n \in N$, failures (demands) are assumed to occur according to a Poisson process with constant rate $m_{i,n}$ (≥ 0). If SKU i does not occur in the material breakdown structures of the machines in group n, then $m_{i,n} = 0$ by definition. Let $M_n := \sum_{i \in I} m_{i,n}$, $n \in N$, and assume that $M_n > 0$, $n \in N$.

If one of the parts of a machine fails, the machine is down and the defective part has to be replaced by a spare part. A failure of a machine is always caused by one defective part, and can be remedied by replacing that part only. All requests for spare parts are sent to the warehouse. For machine group $n \in N$, it is required that the aggregate mean waiting time is at most equal to $\hat{W}_n^{\text{obj}}$ (> 0). We assume that $\hat{W}_1^{\text{obj}} \leq \ldots \leq \hat{W}_{|N|}^{\text{obj}}$.

For each SKU $i \in I$, the stock in the warehouse is controlled by a continuous-review *critical level policy*. This means that the total stock for SKU i is controlled by a basestock policy with basestock level S_i $(\in \mathbb{N}_0 := \mathbb{N} \cup \{0\})$, and that there is a critical level $h_{i,n}$ $(\in \mathbb{N}_0)$ per group $n \in N$, with $h_{i,1} \leq \ldots \leq h_{i,|N|} \leq S_i$, $i \in I$ (we refer to this ordering for the critical levels as the *monotonicity constraint*). A critical level policy for SKU i is denoted by vector $(\mathbf{h}_i, S_i)$, with $\mathbf{h}_i := (h_{i,1}, \ldots, h_{i,|N|})$. If group n demands a part at a moment that the on-hand stock of SKU i is larger than $h_{i,n}$, then this demand is satisfied immediately from the stock in the warehouse. Otherwise, the demand is fulfilled from another source that sends the required spare part to the customer via an emergency shipment. For the warehouse, this demand can be considered as a lost sale. The average time for an emergency shipment is t_i^{em} (≥ 0) and the corresponding cost is c_i^{em} (≥ 0). The cost parameter c_i^{em} models the *extra* costs that one has in comparison to a regular supply of a part of SKU i. We set $h_{i,1} := 0$, since it is not logical to deny group 1 a spare part. (For a single-item problem related to this problem, this has been shown to be optimal; see Lemma 2 of [14].) A policy for all SKU's is denoted by $(\mathbf{H}, \mathbf{S})$, where the matrix

$$\mathbf{H} = \begin{pmatrix} h_{1,1} & h_{1,2} & \ldots & h_{1,|N|} \\ h_{2,1} & h_{2,2} & \ldots & h_{2,|N|} \\ \vdots & \vdots & \ddots & \vdots \\ h_{|I|,1} & h_{|I|,2} & \ldots & h_{|I|,|N|} \end{pmatrix}$$

contains all critical levels and the vector $\mathbf{S} = (S_1, \ldots, S_{|I|})$ contains all basestock levels.

When a part of SKU i in the warehouse stock is used to fulfill a demand, a ready-for-use part of SKU i arrives in the warehouse to refill the stock after a regular repair (or replenishment) leadtime with mean t_i $(> t_i^{\text{em}})$. Per SKU, lead times are independent and identically distributed, and lead times for different SKU's are independent.

Let $\beta_{i,n}(\mathbf{h}_i, S_i)$ denote the fill rate for SKU $i \in I$ and group $n \in N$ under critical level policy $(\mathbf{h}_i, S_i)$, i.e., the percentage of requests for SKU i by group n that can be delivered immediately from the warehouse. An expression for $\beta_{i,n}(\mathbf{h}_i, S_i)$ can be derived as follows. If the number of parts of SKU i in the replenishment pipeline is $k \in \{0, \ldots, S_i\}$, and thus the number of parts in the on-hand stock is $S_i - k$, then the demand rate equals $\mu_{i,k} = \sum_{n|k<(S_i - h_{i,n})} m_{i,n}$, $k \in \{0, \ldots, S_i - 1\}$. For a given SKU i, our inventory model can be described by a closed queueing network with S_i customers and two stations: (i) an ample server with mean service time t_i, which represents the pipeline stock; (ii) a load-dependent, exponential, single server with first-come first-served service discipline, which represents the on-hand stock.

The service rates of the load-dependent server are given by $\mu_{i,k}$. If service times at the ample server would be Phase-Type distributed, this closed queueing network would satisfy the definition of a so-called BCMP-network, see Baskett et al. [3]. In 1976, Barbour [2] proved that, for symmetric queues, one can even allow general service times (for a definition of symmetric queues, see e.g. pp. 338–340 of [23]). Since the queue at the ample server is symmetric, the results of Baskett et al. hold for our closed queueing network. According to their theorem (pp. 253–254 of [3]), the steady-state probability $q_{i,k}$, $k \in \{0,,\dots,S_i\}$, for having k parts in the pipeline is given by

$$q_{i,k} = \left\{\prod_{j=0}^{k-1} \mu_{i,j}\right\} \frac{t_i^k}{k!} q_{i,0}, \quad k \in \{0,\dots,S_i\}, \tag{4.1}$$

$$q_{i,0} = \left(\sum_{k=0}^{S_i} \left\{\prod_{j=0}^{k-1} \mu_{i,j}\right\} \frac{t_i^k}{k!}\right)^{-1}, \tag{4.2}$$

with the convention that $\prod_{j=0}^{k-1} \mu_{i,j} = 1$ for $k = 0$ (this result also follows from [11], pp. 250–252). By these steady state probabilities, we obtain:

$$\beta_{i,n}(\mathbf{h}_i, S_i) = \sum_{k=0}^{S_i - h_{i,n} - 1} q_{i,k}, \quad n \in N, \tag{4.3}$$

with the convention that this sum is empty if $S_i - h_{i,n} - 1 < 0$ (i.e., $\beta_{i,n}(\mathbf{h}_i, S_i) = 0$ if $h_{i,n} = S_i$). Notice that $1 \geq \beta_{i,1}(\mathbf{h}_i, S_i) \geq \ldots \geq \beta_{i,|N|}(\mathbf{h}_i, S_i) \geq 0$.

Further, let $W_{i,n}(\mathbf{h}_i, S_i)$, $i \in I$, $n \in N$, denote the mean waiting time for a request from group n for SKU i. Then:

$$W_{i,n}(\mathbf{h}_i, S_i) = (1 - \beta_{i,n}(\mathbf{h}_i, S_i))t_i^{\mathrm{em}}, \quad i \in I, \; n \in N. \tag{4.4}$$

The aggregate mean waiting time $\hat{W}_n(\mathbf{H}, \mathbf{S})$ for machine group n is equal to

$$\hat{W}_n(\mathbf{H}, \mathbf{S}) = \sum_{i \in I} \frac{m_{i,n}}{M_n} W_{i,n}(\mathbf{h}_i, S_i), \quad n \in N.$$

For each SKU i, $i \in I$, the average costs $C_i(\mathbf{h}_i, S_i)$ consist of inventory holding costs and costs for emergency shipments. The holding cost per time unit per part of SKU i is c_i^{h} (> 0), and applies for the on-hand stock and the parts in the repair pipeline. Hence the inventory holding costs are $c_i^{\mathrm{h}} S_i$. For each machine group n, the number of emergency shipments per time unit equals $m_{i,n}(1 - \beta_{i,n}(\mathbf{h}_i, S_i))$, and thus the emergency shipment costs are equal to $m_{i,n}(1 - \beta_{i,n}(\mathbf{h}_i, S_i))c_i^{\mathrm{em}}$. Hence,

$$C_i(\mathbf{h}_i, S_i) = c_i^{\mathrm{h}} S_i + \sum_{n \in N} m_{i,n}(1 - \beta_{i,n}(\mathbf{h}_i, S_i))c_i^{\mathrm{em}}. \tag{4.5}$$

The total average costs are equal to $C(\mathbf{H}, \mathbf{S}) = \sum_{i \in I} C_i(\mathbf{h}_i, S_i)$.

The objective is to minimize the total average costs subject to the aggregate mean waiting time constraints for the machine groups:

$$\text{(P)} \qquad \min \qquad C(\mathbf{H},\mathbf{S})$$

$$\text{subject to } \hat{W}_n(\mathbf{H},\mathbf{S}) \leq \hat{W}_n^{\text{obj}}, \quad n \in N,$$

$$(\mathbf{H},\mathbf{S}) \in \mathscr{S},$$

where

$$\mathscr{S} = \{(\mathbf{H},\mathbf{S}) \mid h_{i,n}, S_i \in \mathbb{N}_0 \text{ for all } i \in I, n \in N, \\ \text{and } 0 = h_{i,1} \leq h_{i,2} \leq \ldots \leq h_{i,|N|} \leq S_i \text{ for all } i \in I\}.$$

The optimal costs of Problem (P) are denoted by C_{P}.

Via the critical levels policies of our model, we can create customer differentiation. For the situation without customer differentiation, we will use pure basestock policies (i.e., without critical levels), and that situation can then be analyzed by the model of Chap. 3.

4.3 Underlying Single Item Problem

In this section, we consider a single item problem that needs to be solved as a subproblem within the Dantzig-Wolfe decomposition. In this problem, we have a general penalty cost per machine group for demands that are satisfied by an emergency shipment, and the objective is to minimize the total costs within the class of critical level policies. This problem is defined in Sect. 4.3.1. Next, an exact solution under a given, fixed basestock level is described in Sect. 4.3.2. Finally, in Sect. 4.3.3, based on the procedure of Sect. 4.3.2, we describe an exact solution procedure for the full problem of Sect. 4.3.1.

In this section, we follow the same notation as in Sect. 4.2, but we drop the indices for the SKU's $i \in I$.

4.3.1 Single Item Model

We consider a single SKU that is kept on stock in a single warehouse to serve demands of multiple machine groups. The set of machine groups is denoted by N. These groups are numbered $1, \ldots, |N|$ and this numbering is in decreasing order of their importance. The demand process of machine group $n \in N$ is a Poisson process with rate m_n (≥ 0); we assume that $M := \sum_{n \in N} m_n > 0$.

The stock is controlled by a critical level policy $(\mathbf{h},S)$, where $\mathbf{h}=(h_1,\ldots,h_{|N|})$ is the vector of all critical levels and S is the basestock level. For each fulfilled demand, we receive a ready-for-use part after a repair leadtime with mean t. These leadtimes are independent and identically distributed. The critical levels determine when emergency shipments are applied. Let c_n^{p} (≥ 0) denote the penalty cost for each demand of machine group n that is satisfied by an emergency shipment. Because the machine groups are numbered according to their importance, we assume that $c_1^{\mathrm{p}} \geq c_2^{\mathrm{p}} \geq \ldots \geq c_{|N|}^{\mathrm{p}}$ and $h_1 \leq h_2 \leq \ldots \leq h_{|N|}$ (see Remark 4.1 for what to do if the monotonicity for the c_n^{p} does not hold). For machine group 1, there is no reason to satisfy a demand by an emergency shipment as long as there is a part on stock, and therefore we set $h_1 := 0$ (if we would allow $h_1 \geq 0$, then we would obtain $h_1 = 0$ under each optimal solution; see Lemma 2 of [14]). Apart from penalty costs for emergency shipments, we have an inventory holding cost rate c^{h} (> 0) per part on stock or in the repair pipeline. So, the average holding costs are $c^{\mathrm{h}}S$.

To determine the average penalty costs, we need the fill rate $\beta_n(\mathbf{h},S)$ per machine group n, and these fill rates are obtained via the reasoning that we already gave in Sect. 4.2 (see the reasoning leading to Eqs. (4.1)–(4.3)). Let q_k be the steady-state probability for having k parts in the pipeline. Define $\mu_k := \sum_{n|k<(S-h_n)} m_n$, $k \in \{0,\ldots,S-1\}$. Then

$$q_k = \left\{\prod_{j=0}^{k-1}\mu_j\right\}\frac{t^k}{k!}q_0,\quad k\in\{1,\ldots,S\},$$

$$q_0 = \left(\sum_{k=0}^{S}\left\{\prod_{j=0}^{k-1}\mu_j\right\}\frac{t^k}{k!}\right)^{-1},$$

with the convention that $\prod_{j=0}^{k-1}\mu_j = 1$ for $k=0$, and

$$\beta_n(\mathbf{h},S) = \sum_{k=0}^{S-h_n-1} q_k,\quad n\in N,$$

with the convention that this sum is empty if $S-h_n-1<0$ (i.e., $\beta_n(\mathbf{h},S)=0$ if $h_n=S$). It holds that $1\geq\beta_1(\mathbf{h},S)\geq\ldots\geq\beta_{|N|}(\mathbf{h},S)\geq 0$.

The total average costs $\hat{C}(\mathbf{h},S)$ of a critical level policy $(\mathbf{h},S)$ are equal to

$$\hat{C}(\mathbf{h},S) = c^{\mathrm{h}}S + \sum_{n\in N} c_n^{\mathrm{p}} m_n(1-\beta_n(\mathbf{h},S)) \tag{4.6}$$

(we use a hat for $\hat{C}(\mathbf{h},S)$ in order to distinguish this function from the total costs function $C(\mathbf{H},\mathbf{S})$ for the multi-item model of Sect. 4.2). The objective is to minimize these total average costs within the class of critical level policies:

$$
\begin{aligned}
\text{(Q)} \qquad \min \qquad & \hat{C}(\mathbf{h},S) \\
\text{subject to } & 0 = h_1 \leq h_2 \leq \ldots \leq h_{|N|} \leq S, \\
& h_n \in \mathbb{N}_0,\ n \in N, \\
& S \in \mathbb{N}_0.
\end{aligned}
$$

Let $(\mathbf{h}^*, S^*)$ be an optimal policy for Problem (Q) and let $\hat{C}_{\mathrm{Q}} = \hat{C}(\mathbf{h}^*, S^*)$ denote the optimal costs.

For the situation with a given basestock level $S \in \mathbb{N}_0$, let Problem (Q(S)) denote the problem of finding the critical levels such that the average costs $\hat{C}(\mathbf{h},S)$ are minimized:

$$
\begin{aligned}
\text{(Q}(S)) \qquad \min \qquad & \hat{C}(\mathbf{h},S) \\
\text{subject to } & 0 = h_1 \leq h_2 \leq \ldots \leq h_{|N|} \leq S, \\
& h_n \in \mathbb{N}_0,\ n \in N.
\end{aligned}
$$

An optimal solution for Problem (Q(S)) is denoted by $\mathbf{h}^*(S)$ and the corresponding optimal costs are denoted by $\hat{C}_{\mathrm{Q}(S)} = \hat{C}(\mathbf{h}^*(S), S)$. Obviously, $\hat{C}(\mathbf{h}^*(S^*), S^*) = \hat{C}(\mathbf{h}^*, S^*)$, and equivalently $\hat{C}_{\mathrm{Q}(S^*)} = \hat{C}_{\mathrm{Q}}$.

4.3.2 Exact Solution for Problem (Q(S))

In this section, we assume that $S \in \mathbb{N}_0$ is given and we consider Problem (Q(S)). The solution space of Problem (Q(S)) consists of all possible vectors $\mathbf{h} = (h_1, \ldots, h_{|N|})$ with

$$0 = h_1 \leq h_2 \leq \ldots \leq h_{|N|} \leq S. \tag{4.7}$$

We denote this solution space by $\mathscr{S}'$. One way to solve Problem (Q(S)) is enumeration. It can be shown that the total number of solutions in $\mathscr{S}'$ is equal to (this number is equal to the number of possibilities to divide S balls over $|N|$ buckets)

$$\binom{S+|N|-1}{|N|-1}.$$

This number becomes large for large values of the basestock level S and the number of machine groups $|N|$. Hence, it is desired to have a faster solution method. By advanced results of Van Jaarsveld [21], we can get a fast and exact solution by a local search algorithm. This is described below.

For each $\mathbf{h} \in \mathscr{S}'$, we define its neighborhood $\mathscr{N}(\mathbf{h})$ by

$$\mathscr{N}(\mathbf{h}) := \{\mathbf{h}' \in \mathscr{S}' | \mathbf{h}' = \mathbf{h} + \mathbf{e}_n,\ n \in N,\ n \geq 2\} \\ \cup \{\mathbf{h}' \in \mathscr{S}' | \mathbf{h}' = \mathbf{h} - \mathbf{e}_n,\ n \in N,\ n \geq 2\},$$

where $\mathbf{e}_n$ is an $|N|$-dimensional unit vector with a 1 on the n-th position. The neighbors of a solution $\mathbf{h}$ are other solutions that one can obtain by increasing or decreasing one critical level h_n with one unit without violating (4.7). Notice that the number of neighbors of $\mathbf{h}$ is at most equal to $2^{|N|-1}$. A solution $\mathbf{h} \in \mathscr{S}'$ for which all neighbors have equal or strictly larger costs is called a *locally optimal* solution. Based on this neighborhood definition, we obtain the following local search algorithm, which gives us such a locally optimal solution.

Algorithm 4.1 *(Local search algorithm for Problem (Q(S)))*

Step 1 Set $\mathbf{h} := (0,\ldots,0)$ and determine $\hat{C}(\mathbf{h},S)$.
Step 2 Determine $\hat{C}(\mathbf{h}',S)$ for all $\mathbf{h}' \in \mathscr{N}(\mathbf{h})$;
Select the $\mathbf{h}'$ with the lowest average costs $\hat{C}(\mathbf{h}',S)$.
Step 3 If $\hat{C}(\mathbf{h}',S) < \hat{C}(\mathbf{h},S)$, then $\mathbf{h} := \mathbf{h}'$ and goto Step 2, else stop.

Because the set $\mathscr{S}'$ has a finite number of solutions, and we only jump to a neighbor when that neighbor has strictly lower costs, we know that Algorithm 4.1 terminates after a finite number of steps.

By a result of Van Jaarsveld [21], any locally optimal solution (under the given neighborhood structure) is also globally optimal, i.e., it is an optimal solution for Problem (Q(S)). This immediately implies that Algorithm 4.1 leads to an optimal solution for Problem (Q(S)). These results are stated in the following theorem and corollary.

Theorem 4.1. *A locally optimal solution* $\mathbf{h} \in \mathscr{S}'$ *is an optimal solution for Problem (Q(S)).*

Proof. Let us first look at the special case with an exponential distribution with mean t for the replenishment leadtime. Then our theorem follows from Lemma 7.6 of Van Jaarsveld [21]. In the model in [21] to which Lemma 7.6 applies, it is allowed that $h_1 \geq 0$, while we limit ourselves to solutions with $h_1 = 0$. Hence, we have to add one additional step to the proof, which goes as follows.

Let $\mathbf{h} \in \mathscr{S}'$ be a locally optimal solution of our model. If $h_1 < h_2$, then one can show that $\hat{C}(\mathbf{h},S) \leq \hat{C}(\mathbf{h}+\mathbf{e}_1,S)$; this can be shown via the reasoning in the proof of Lemma 2 in [14]. Next we apply Lemma 7.6 of [21], which implies that $\mathbf{h}$ is globally optimal for the model of [21] and hence also for our Problem (Q(S)). (In [21], one studies a wider class of policies. In fact, all possible policies are allowed and it is also shown that the class of critical policies is optimal under a given fixed basestock level S.)

Let us now move to a general leadtime distribution with mean t for the replenishment leadtime. Because the average costs $\hat{C}(\mathbf{h},S)$ of any solution $\mathbf{h}$ only depend on the leadtime distribution via the mean t, it is trivial that then our theorem also applies. □

Corollary 4.1. *The solution obtained by Algorithm 4.1 is an optimal solution of Problem (Q(S)).*

Finally, we would like to comment on the choice $\mathbf{h} := (0,\ldots,0)$ for the starting solution of Algorithm 4.1. We may expect that an optimal solution $\mathbf{h}^*(S)$ has low critical levels for most problem instances. Hence, by starting with $\mathbf{h} := (0,\ldots,0)$, we will generally reach an optimal solution within a limited number of steps.

4.3.3 Exact Solution for Problem (Q)

In this section, we describe an exact method to solve Problem (Q). This method is based on convex lower and upper bound functions for the function $\hat{C}(\mathbf{h}^*(S),S)$ (cf. the approach for the equivalent problem in [7]).

Define

$$\tilde{C}(S) := \hat{C}(\mathbf{h}^*(S),S), \;\; S \in \mathbb{N}_0.$$

For a given $S \in \mathbb{N}_0$, an upper bound is obtained by taking all critical levels equal to 0. This gives the upper bound $\tilde{C}^{\text{ub}}(S) = \hat{C}(\mathbf{0},S)$, where $\mathbf{0} = (0,\ldots,0)$ is an $|N|$-dimensional vector with zeros on all positions. When all critical levels are zero, the behavior of the inventory is the same as for a stockpoint with one aggregated machine group with demand rate M. Let $\beta(S)$ be the fill rate for the latter system. The behavior of this latter system is equivalent to the behavior of an Erlang loss system (i.e., an $M|G|c|c$ queue) with S servers, arrival rate M, and mean service time t (see also Sect. 2.9), and hence $\beta(S)$ is equal to 1 minus the corresponding Erlang loss probability:

$$\beta(S) = 1 - \frac{\frac{1}{S!}(Mt)^S}{\sum_{j=0}^{S} \frac{1}{j!}(Mt)^j}.$$

It holds that, for each machine group n, its fill rate $\beta_n(\mathbf{0},S)$ is equal to this fill rate $\beta(S)$, and hence

$$\begin{aligned}\tilde{C}^{\text{ub}}(S) &= \hat{C}(\mathbf{0},S) = c^{\text{h}}S + \sum_{n\in N} c^{\text{p}}_n m_n (1-\beta_n(\mathbf{0},S)) \\ &= c^{\text{h}}S + \left(\sum_{n\in N} c^{\text{p}}_n m_n\right)(1-\beta(S)). \qquad (4.8)\end{aligned}$$

For a given $S \in \mathbb{N}_0$, a lower bound is obtained by replacing all penalty cost parameters c^{p}_n by the smallest penalty cost parameter $c^{\text{p}}_{|N|}$. This defines the lower bound $\tilde{C}^{\text{lb}}(S)$. Under these modified cost parameters, all machine groups have the same penalty cost parameter and thus it is optimal to choose all critical levels equal to 0.

The resulting system behaves as a stockpoint with one aggregated machine group with demand rate M, and we find

$$\begin{aligned}\tilde{C}^{\text{lb}}(S) &= c^{\text{h}}S + \sum_{n\in N} c^{\text{p}}_{|N|} m_n(1-\beta_n(\mathbf{0},S)) \\ &= c^{\text{h}}S + c^{\text{p}}_{|N|} M(1-\beta(S)).\end{aligned} \tag{4.9}$$

The term $1-\beta(S)$ in (4.8) and (4.9) equals the Erlang loss probability for the Erlang loss system mentioned above and this probability is known to be strictly convex and decreasing; see [13] (see also Remark 2 in [14]). Hence, both the upper bound function $\tilde{C}^{\text{ub}}(S)$ and lower bound function $\tilde{C}^{\text{lb}}(S)$ is convex on the entire domain $S \in \mathbb{N}_0$. As a result, we get the following exact solution procedure for Problem (Q).

First, define S' as a minimizing point for the upper bound function $\tilde{C}^{\text{ub}}(S)$. Next, S^{lb} is defined as the lowest $S \in \mathbb{N}_0$ for which $\tilde{C}^{\text{lb}}(S) \leq \tilde{C}^{\text{ub}}(S')$. This S^{lb} is a lower bound for the optimal base stock level S^*. Then, we solve Problem (Q(S)) for $S = S^{\text{lb}}, S^{\text{lb}}+1, S^{\text{lb}}+2, \ldots$ by Algorithm 4.1. While doing so, we keep track of the best solution found until a certain point, and we denote this solution by $(\mathbf{h}^*(S^{\text{best}}), S^{\text{best}})$. After having solved Problem (Q(S)) for a given S and having updated $(\mathbf{h}^*(S^{\text{best}}), S^{\text{best}})$ if a better solution was found for this S, we can guarantee that we have found the optimal solution if the following two conditions are satisfied: (i) $\tilde{C}^{\text{lb}}(S+1) \geq \tilde{C}^{\text{lb}}(S)$; (ii) $\tilde{C}(S^{\text{best}}) \leq \tilde{C}^{\text{lb}}(S+1)$. Condition (i) implies that the lower bound function is increasing from this S on. Hence, when both conditions are satisfied, we know that for any base stock level larger than S no better solution can be found than the current best solution. Because $\tilde{C}^{\text{lb}}(S) \to \infty$ when $S \to \infty$, we know that we will satisfy the conditions (i) and (ii) at some finite S (at the latest at the largest S where $\tilde{C}^{\text{lb}}(S) \leq \tilde{C}^{\text{ub}}(S')$). This procedure is formally described as follows.

Algorithm 4.2 *(Exact solution procedure for Problem (Q))*

Step 1 Determine $S' := \arg\min \tilde{C}^{\text{ub}}(S)$;
Determine $S^{\text{lb}} := \min\{S | \tilde{C}^{\text{lb}}(S) \leq \tilde{C}^{\text{ub}}(S')\}$;
$S := S^{\text{lb}}$ and $S^{\text{best}} := S$.
Step 2 Solve Problem (Q(S)) by Algorithm 4.1, by which one obtains $\mathbf{h}^*(S)$ and $\tilde{C}(S)$;
If $\tilde{C}(S) < \tilde{C}(S^{\text{best}})$, then $S^{\text{best}} := S$.
Step 3 If $\tilde{C}^{\text{lb}}(S+1) \geq \tilde{C}^{\text{lb}}(S)$ and $\tilde{C}(S^{\text{best}}) \leq \tilde{C}^{\text{lb}}(S+1)$,
then stop and goto Step 4, else $S := S+1$ and goto Step 2.
Step 4 $S^* := S^{\text{best}}$ and $\mathbf{h}^* := \mathbf{h}^*(S^{\text{best}})$.

Remark 4.1. In the model as described above, it is assumed that penalty cost parameters c^{p}_n, $n \in N$, satisfy $c^{\text{p}}_1 \geq \ldots \geq c^{\text{p}}_{|N|}$. In the subproblems that need to be solved in the Dantzig-Wolfe decomposition (in Sect. 4.4), this monotonicity property does not always hold, while we do require that $h_1 \leq h_2 \leq \ldots \leq h_{|N|}$. In that case, we may proceed as follows. Suppose we have two succeeding groups n and $n+1$, $n < |N|$, with $c^{\text{p}}_n \leq c^{\text{p}}_{n+1}$. Then it is never optimal to favor group n above $n+1$ and thus, under the monotonicity constraint saying that $h_n \leq h_{n+1}$, it is best to have $h_n = h_{n+1}$

(i.e., this will happen under an optimal solution). As a result, we can merge such two groups. For this merged group, the demand rate is equal to the sum of the two original demand rates ($= m_n + m_{n+1}$), and the penalty cost parameter becomes equal to the demand-weighted average of the penalty cost parameters of the two original groups, i.e., equal to

$$\frac{m_n}{m_n + m_{n+1}} c_n^{\mathrm{p}} + \frac{m_{n+1}}{m_n + m_{n+1}} c_{n+1}^{\mathrm{p}}.$$

After this transformation, the monotonicity property may hold. Otherwise, one or more transformation steps can be executed until the monotonicity property does hold. Once the monotonicity property holds, the problem may be solved by the exact solution method (Algorithm 4.2).

4.4 Dantzig-Wolfe Decomposition

In this section we return to the multi-item problem of Sect. 4.2 and we describe the application of Dantzig-Wolfe decomposition (cf. [6]). This leads to a lower bound for the optimal costs C_{P} (see Sect. 4.4.1) and a heuristic solution (see Sect. 4.4.2).

4.4.1 Lower Bound

We first introduce a *Master Problem*, in which the variables of our original problem are expressed as convex combinations of columns that contain all possible values for the decision variables in the original problem. Let K denote the set of critical level policies for each of the SKU's $i \in I$, satisfying $0 = h_{i,1} \leq \ldots \leq h_{i,|N|} \leq S_i$, $h_{i,n} \in \mathbb{N}_0$, $n \in N$, and $S_i \in \mathbb{N}_0$. Let vector $(\mathbf{h}_i^k, S_i^k)$, $i \in I, k \in K$, with $\mathbf{h}_i^k := (h_{i,1}^k, \ldots, h_{i,|N|}^k)$, denote the critical levels and basestock level of policy k for SKU i, and let $x_i^k \in \{0,1\}$, $i \in I, k \in K$, be a variable indicating whether policy k for SKU i is chosen ($x_i^k = 1$) or not ($x_i^k = 0$). Relaxing the integrality constraint on x_i^k, $i \in I, k \in K$, a suitable Master Problem related to Problem (P) is defined as follows:

$$\text{(MP)} \qquad \min \quad \sum_{i \in I} \sum_{k \in K} C_i(\mathbf{h}_i^k, S_i^k) x_i^k$$

$$\text{subject to} \sum_{i \in I} \sum_{k \in K} \frac{m_{i,n}}{M_n} W_{i,n}(\mathbf{h}_i^k, S_i^k) x_i^k \leq \hat{W}_n^{\mathrm{obj}}, \quad n \in N, \qquad \text{(MP.1)}$$

$$\sum_{k \in K} x_i^k = 1, \qquad i \in I, \qquad \text{(MP.2)}$$

$$x_i^k \geq 0, \qquad i \in I, k \in K.$$

The optimal costs of Problem (MP) are denoted by C_{MP}. Notice that the relaxation of the integrality condition on x_i^k, $i \in I, k \in K$, in Problem (MP) allows for fractional values of x_i^k, $i \in I, k \in K$, and thus corresponds to allowing *randomized policies* (also called *mixed policies*, see e.g. [19], Sect. 8.9.2). Therefore, C_{MP} constitutes a lower bound for C_{P}.

Besides Problem (MP), a Restricted Master Problem (RMP) is defined. In this restricted problem, we only consider a small subset $K_i \subseteq K$ of columns (policies) for each SKU $i \in I$. The optimal costs of Problem (RMP) are denoted by C_{RMP}. For each SKU $i \in I$, let K_i initially consist of one policy k, with all critical levels equal to zero and $S_i^k := \min\{S_i \in \mathbb{N}_0 | W_{i,n}((0,\ldots,0),S_i) \leq \hat{W}_n^{\mathrm{obj}} \text{ for all } n \in N\}$. This choice of initial policies constitutes a feasible solution for Problem (RMP). That S_i^k is finite for each $i \in I$, can be seen as follows. First, it holds that $\beta_{i,n}((0,\ldots,0),S_i)$ is strictly increasing on its entire domain $\mathbb{N}_0$ (like the fill rate $\beta(S)$ as discussed in Sect. 4.3.3). Hence $W_{i,n}((0,\ldots,0),S_i)$ is strictly decreasing on its entire domain $\mathbb{N}_0$. Furthermore, $\beta_{i,n}((0,\ldots,0),\infty) = 1$, and thus $W_{i,n}((0,\ldots,0),\infty) = 0$. So, there is a finite S_i for which $W_{i,n}((0,\ldots,0),S_i) \leq \hat{W}_n^{\mathrm{obj}}$ for each $n \in N$, and thus S_i^k is finite.

After solving Problem (RMP) with the $|I|$ initial policies by using the simplex method, we are interested in policies that were not yet considered, but that would improve the solution of Problem (RMP) if they were added. To check whether such policies exist, we solve, for each SKU $i \in I$, a so-called *column generation subproblem* that for SKU $i \in I$ generates a policy with the lowest reduced cost coefficient. Given an optimal solution for Problem (RMP), let $u_n \leq 0$, $n \in N$, denote the dual variables (shadow prices) related to the $|N|$ waiting time constraints (MP.1), and let v_i, $i \in I$, denote the dual variables related to the $|I|$ constraints (MP.2). Let $C_i^{\mathrm{red}}(\mathbf{h}_i, S_i)$ denote the reduced cost of a policy $(\mathbf{h}_i, S_i)$ for SKU $i \in I$, i.e.,

$$
\begin{aligned}
C_i^{\mathrm{red}}(\mathbf{h}_i, S_i) &= C_i(\mathbf{h}_i, S_i) - \sum_{n \in N} u_n \frac{m_{i,n}}{M_n} W_{i,n}(\mathbf{h}_i, S_i) - v_i \\
&= c_i^{\mathrm{h}} S_i - v_i + \sum_{n \in N} \left(c_i^{\mathrm{em}} - \frac{u_n t_i^{\mathrm{em}}}{M_n} \right) m_{i,n} (1 - \beta_{i,n}(\mathbf{h}_i, S_i)).
\end{aligned}
$$

Then, for our Problem (RMP), the column generation subproblem for SKU i, is as follows.

$$
\begin{aligned}
(\mathrm{SUB}(i)) \min \quad & C_i^{\mathrm{red}}(\mathbf{h}_i, S_i) \\
\text{subject to } & 0 = h_{i,1} \leq \ldots \leq h_{i,|N|} \leq S_i, \\
& h_{i,n},\ S_i \in \mathbb{N}_0,\ n \in N.
\end{aligned}
$$

Let $(\mathbf{h}_{SUB(i)}, S_{SUB(i)})$ denote an optimal policy (critical levels and basestock level) for Problem (SUB(i)), and let $C_{\mathrm{SUB}(i)}$ denote the cost of an optimal solution of Problem (SUB(i)), i.e., the lowest reduced cost coefficient, which also can be interpreted as the degree of violation of the corresponding constraint in the dual problem of Problem (MP).

Problem (SUB(i)) is equivalent to Problem (Q) as defined in Sect. 4.3.1. This equivalence is as follows. Replace all variables in the model of Sect. 4.3.1 by the variables of SKU i, which includes replacing the critical levels vector $\mathbf{h}$ by $\mathbf{h}_i$, the basestock level S by S_i and the penalty cost parameters c_n^{p} by the factors $c_i^{\mathrm{em}} - (u_n t_i^{\mathrm{em}})/M_n$. Then

$$\hat{C}(\mathbf{h}, S) = C_i^{\mathrm{red}}(\mathbf{h}_i, S_i) + v_i,$$

i.e., the objective function of Problem (Q) is equal to the objective function of Problem (SUB(i)) plus a constant term v_i. Obviously, this constant term has no effect on the optimal solution that is obtained. Further the constraints are identical. Hence, we can solve Problem (SUB(i)) by the exact solution method for Problem (Q) (i.e., by Algorithm 4.2). Notice that the penalty cost parameters $c_n^{\mathrm{p}} = c_i^{\mathrm{em}} - (u_n t_i^{\mathrm{em}})/M_n$ will not be monotonic (decreasing for increasing n) in general; this issue can be addressed via the transformation described in Remark 4.1.

As long as a policy with a negative reduced cost coefficient is found for one or more SKU's, adding columns to Problem (RMP) (i.e., policies found as solutions for Problem (SUB(i)) for $i \in I$) and solving Problem (RMP) is done iteratively. If for none of the SKU's $i \in I$ a policy with negative reduced cost can be found, the obtained solution for Problem (RMP) is optimal for Problem (MP) as well.

Regarding the convergence of this iterative procedure, similar properties hold as for the Dantzig-Wolfe decomposition for the problem of Chap. 3; see the last but one paragraph of Sect. 3.3.3.1. In short, it is known that the procedure will convergence in general. Further, we obtained convergence in all instances to which the procedure has been applied.

In total, $|I| + |N|$ variables are in the basis. Notice that at most $|N|$ SKU's will have fractional x_i^k-values because the constraints (MP.2) require that for each i at least one x_i^k is a basic variable. Furthermore, notice that the number of waiting time constraints (MP.1) that is satisfied with equality is at least equal to the number of SKU's that has fractional x_i^k-values, again because of the number of variables in the basis.

4.4.2 Heuristic Solution

After a lower bound has been found for Problem (P), an upper bound is obtained by constructing a feasible solution for Problem (P). This is done as follows. If none of the x_i^k-values in the solution of Problem (MP) is fractional, the obtained solution of Problem (MP) is feasible for Problem (P) as well (and thus optimal). If fractional x_i^k-values occur, however, we need to apply some further steps. Select for each SKU $i \in I$ the policy with a non-zero x_i^k-value and the lowest basestock level S_i^k, and refer to this policy as policy k'. Set $(\mathbf{h}_i, S_i) := (\mathbf{h}_i^{k'}, S_i^{k'})$ for all SKU's $i \in I$. This gives an overall policy $(\mathbf{H}, \mathbf{S})$. If this policy $(\mathbf{H}, \mathbf{S})$ constitutes a feasible solution for Problem (P), we are ready. Otherwise, we increase the basestock levels in a greedy way until we obtain a feasible solution (we do not change the critical levels). This greedy way works as follows.

For each solution $(\mathbf{H},\mathbf{S})$, we define the distance to the feasible solution by

$$d(\mathbf{H},\mathbf{S}) := \sum_{n \in N} \left(\hat{W}_n(\mathbf{H},\mathbf{S}) - \hat{W}_n^{\text{obj}} \right)^+ ,$$

with $x^+ := \max\{0,x\}$ for all $x \in \mathbb{R}$. We use the operator Δ_i to denote the change in the total costs $C(\mathbf{H},\mathbf{S})$ and the distance $d(\mathbf{H},\mathbf{S})$ when the basestock level S_i is increased with one unit. It holds that

$$\begin{aligned} \Delta_i C(\mathbf{H},\mathbf{S}) &= C_i(\mathbf{h}_i, S_i + 1) - C_i(\mathbf{h}_i, S_i), \\ \Delta_i d(\mathbf{H},\mathbf{S}) &= d(\mathbf{H},\mathbf{S} + \hat{\mathbf{e}}_i) - d(\mathbf{H},\mathbf{S}) \\ &= \sum_{n \in N} \left[\left(\hat{W}_n(\mathbf{H},\mathbf{S} + \hat{\mathbf{e}}_i) - \hat{W}_n^{\text{obj}} \right)^+ - \left(\hat{W}_n(\mathbf{H},\mathbf{S}) - \hat{W}_n^{\text{obj}} \right)^+ \right]. \end{aligned}$$

where $\hat{\mathbf{e}}_i$ is an $|I|$-dimensional unit vector with a one on the i-th position. The factor

$$\Gamma_i = \frac{-\Delta_i d(\mathbf{H},\mathbf{S})}{\Delta_i C(\mathbf{H},\mathbf{S})}$$

denotes the decrease in distance to the feasible solutions per unit of increase in costs. For the SKU i with the highest ratio Γ_i, we have the "biggest bang for the buck" and we increase the basestock level to $S_i + 1$. We execute such greedy steps until we arrive at a feasible solution.

Remark 4.2. As mentioned in the introduction, the model of this chapter can also be used for situations with multiple machine types and multiple target levels for the aggregate mean waiting time per machine group. One can then define one machine group per combination of a machine types and a corresponding target level. These machine groups form the set N and they can be ordered such that the corresponding targets $\hat{W}_n^{\text{obj}}$ are increasing as a function of n. In this way, we can directly use the model and analysis of this chapter.

4.5 Computational Experiment

In this section, we execute a computational experiment with a twofold aim. First, we investigate the quality of the Dantzig-Wolfe heuristic, which we measure per instance by the relative distance to the lower bound C_{MP} for the optimal costs C_{P}:

$$G := \frac{C_{\text{DW}} - C_{\text{MP}}}{C_{\text{MP}}},$$

where C_{DW} denotes the costs of the solution generated by the Dantzig-Wolfe heuristic. Second, we compare the costs of our Dantzig-Wolfe solution to the costs of a close-to-optimal solution when no critical levels are being used, i.e., within the class of pure basestock policies. This latter comparison shows the benefits of the use of critical levels.

We consider instances with 2 and 5 machine groups and several settings for the other parameters. The choice of the values for demand rates, target waiting times, inventory holding cost parameters, and transportation costs and times is partly based on what we observed in practice.

Table 4.1 contains the parameter settings for the instances with $|N| = 2$ machine groups, and Table 4.2 for the instances with $|N| = 5$. In all instances, we assume that machines are identical and that they all have the same failure rates for all components. This implies that all machine groups have the same mix of demand rates: For each pair of machine groups n_1 and n_2, it holds that the ratio $m_{i,n_1}/m_{i,n_2}$ is the same for all SKU's $i \in I$. It also implies that the demand rates $m_{i,n}$, $i \in I$, $n \in N$, are fully determined by the total demand rate per SKU i and the ratios between the demand rates for different machine groups. The latter numbers are specified in the Tables 4.1 and 4.2.

Because all machine groups have the same mix of demand rates, the situation when no critical levels are being used becomes equivalent to meeting the lowest target waiting time $\hat{W}_1^{\mathrm{obj}}$ for all machine groups. In the literature, this is known as the *round-up policy* (cf. Deshpande et al. [8]). Meeting the target $\hat{W}_1^{\mathrm{obj}}$ for all groups implies that we can merge all groups and the resulting problem is analyzed as a special case of the model of this chapter.

The further information about our test bed is as follows. First, to obtain base values for the demand, we generate 20 samples of 100 numbers each, randomly drawn from the uniform distribution $U[0,1]$ (we have up to 100 SKU's in our experiment). Also, to obtain base values for the inventory holding cost parameters, we generate 20 samples of 100 numbers each, again randomly drawn from the uniform distribution $U[0,1]$. For each setting of the parameters that we describe below, we generate 20 instances that are based on these base values. Thus, we exclude the effect of coincidence that may occur if we study only one instance, and on the other hand we improve the comparison between the different parameter settings since we use the same random values each time.

In our experiment, we have the following parameter settings. We consider 3 values for the number of SKU's $|I|$: 20, 50, and 100 (in the first two cases, only the first 20, respectively 50 numbers in each sample are used). We consider 3 values for the average inventory holding cost parameter $\overline{c}^{\mathrm{h}}$ and 2 values for the c_i^{h}-range. The latter gives the quotient between the upper and lower limit of the domain for the holding cost values. As an example: For $\overline{c}^{\mathrm{h}} = 10$ and c_i^{h}-range = 9, the holding cost parameters c_i^{h}, $i \in I$, are in the range $[2,18]$, where the precise value for each SKU i is determined as $2 +$ (16 times the base value for the inventory holding cost parameter for that SKU in the sample). The total daily demand rate per SKU can be within 2 ranges, that are obtained by multiplying the base values for the demand by 0.1 and 0.5, respectively.

Table 4.1 Parameter settings for the situation with 2 customer groups ($|N| = 2$)

Parameter	Number of choices	Values
Number of SKU's ($\|I\|$)	3	20, 50, 100
Av. holding cost parameter ($\bar{c}^{\text{h}}$; per day)	3	10, 100, 1,000
Holding cost range (c_i^{h}-range)	2	9, 999
Total demand rate per SKU (per day)	2	U[0, 0.1], U[0, 0.5]
Ratio ($m_{i,1} : m_{i,2}$)	3	$(0.2 : 0.8)$, $(0.5 : 0.5)$, $(0.8 : 0.2)$
Target waiting times ($\hat{W}_n^{\text{obj}}$; in hours)	4	(0.5, 4), (0.5, 8), (1, 4), (1, 8)
Emergency shipment cost (c_i^{em})	1	1,000
Mean emergency shipm. time (t_i^{em}; in days)	1	2
Mean repl. leadtime (t_i; in days)	1	14

Table 4.2 Parameter settings for the situation with 5 customer groups ($|N| = 5$)

Parameter	Number of choices	Values
Number of SKU's ($\|I\|$)	3	20, 50, 100
Av. holding cost parameter ($\bar{c}^{\text{h}}$; per day)	3	10, 100, 1,000
Holding cost range (c_i^{h}-range)	2	9, 999
Total demand rate per SKU (per day)	2	U[0, 0.1], U[0, 0.5]
Target waiting times ($\hat{W}_n^{\text{obj}}$; in hours)	1	(0.5, 1, 2, 4, 8)
Ratio ($m_{i,1} : \ldots : m_{i,5}$)	1	$(0.2 : \ldots : 0.2)$
Emergency shipment cost (c_i^{em})	1	1,000
Mean emergency shipm. time (t_i^{em}; in days)	1	2
Mean repl. lead time (t_i; in days)	1	14

For the instances with $|N| = 2$ customer groups, we vary the ratio of the demand rates of both machine groups (3 choices) and the target waiting times for both groups (4 choices). For $|N| = 5$, we have only one setting for both the ratio of the demand rates and the target waiting times. For both $|N| = 2$ and $|N| = 5$, we have one setting for the t_i, t_i^{em}, c_i^{em}. As a result, we have 8,640 instances with $|N| = 2$, 720 instances with $|N| = 5$, and 9,360 instances in total.

In Table 4.3, we display the results for the instances with 2 customer groups. For various subsets of instances, we list the average and maximum gap G, the average fraction of the SKU's with non-zero critical levels, and the average and maximum cost savings in comparison to the round-up policy. In the bottom row, results are given for all instances together. Other rows give results for subsets of instances that have a parameter setting as indicated. This enables us to study the influence of different parameters. In Table 4.4, results are displayed for the instances with 5 customer groups.

The model has been implemented in AIMMS 3.5, with XA used to solve the LP problems, and the experiment was run on a PC with an Intel Pentium 4, 2.0 GHz processor and 256 MB memory. The average computation time for the instances with 2 groups was around 4 s, and all times were within a minute. The average computation time for the instances with 5 groups was around 1 min and the maximum was below 10 min. The larger computation times are observed for instances with a large total demand rate and a large number of SKU's.

Table 4.3 Results for the situation with two groups ($|N| = 2$)

Parameter		G (%)		SKU's with non-zero critical levels (%)	Savings (%)	
		Average	Maximum	Average	Average	Maximum
$\|I\|$	20	0.76	5.88	67.15	8.02	31.44
	50	0.25	1.70	68.20	8.34	29.34
	100	0.10	0.87	68.42	8.43	29.72
$\bar{c}^{\mathrm{h}}$	10	0.16	2.34	47.40	2.02	9.71
	100	0.47	5.71	78.20	10.76	26.79
	1,000	0.49	5.88	78.18	12.02	31.44
c_i^{h}-range	9	0.36	5.08	68.84	8.11	29.07
	999	0.38	5.88	67.02	8.43	31.44
$\sum_{n\in N} m_{i,n}$	$U[0,0.1]$	0.51	5.88	64.82	7.85	26.63
	$U[0,0.5]$	0.23	2.38	71.04	8.69	31.44
$(m_{i,1} : m_{i,2})$	$(0.2 : 0.8)$	0.43	5.83	55.25	13.39	31.44
	$(0.5 : 0.5)$	0.36	5.88	70.58	8.23	18.65
	$(0.8 : 0.2)$	0.32	3.95	77.95	3.18	7.52
$\hat{W}_n^{\mathrm{obj}}$	(0.5, 4)	0.34	3.69	75.81	7.77	20.81
	(0.5, 8)	0.37	3.93	86.05	10.98	31.44
	(1, 4)	0.37	5.83	43.38	5.27	15.44
	(1, 8)	0.40	5.88	66.46	9.05	27.05
All instances		0.37	5.88	67.93	8.27	31.44

Tables 4.3 and 4.4 show that both the average and maximum relative difference G decreases significantly if the number of SKU's increases. This is an important observation because one generally has large numbers of SKU's in practice. Furthermore, the average G is decreasing for decreasing values of the average inventory holding cost parameter and increasing values of the total demand rate. On average, G is smaller than 1 %. This implies that the heuristic solutions are good.

To determine the savings, we compare the costs of our heuristic solution, C_{DW}, to the costs of the round-up policy for the situation with one aggregated customer group. Obviously, the generated round-up policies are also heuristic solutions. But, they appear to have very small gaps with the lower bound: on average less than 0.2 %, and the maximum gap was 1.8 %.

On average, for more than two-thirds of the SKU's non-zero critical levels are used, and thus differentiated service levels are offered to different machine groups. We examined some instances in further detail, and we observed that typically the SKU's with both high failure rates and high inventory holding cost parameters are the first to have non-zero critical levels, as can be seen for one data set in Fig. 4.1. In spare parts environments, such SKU's that are expensive and break down often are the most likely candidates to be revised by the development department; thus these SKU's do not occur often in real-life data sets. Nevertheless, in the real-life data sets in the case study, we observe a similar behavior. Also then expensive SKU's with high demand rates are the first to have non-zero critical levels (see Figs. 4.2 and 4.3).

Table 4.4 Results for the situation with five groups ($|N| = 5$)

Parameter		G (%)		SKU's with non-zero critical levels (%)	Savings (%)	
		Average	Maximum	Average	Average	Maximum
$\|I\|$	20	1.91	7.95	82.94	8.51	16.27
	50	0.72	2.46	84.95	9.48	15.63
	100	0.34	1.23	85.45	9.86	15.93
$\bar{c}^{\mathrm{h}}$	10	0.58	3.58	63.33	3.34	5.77
	100	1.17	7.95	95.01	11.68	14.38
	1,000	1.23	5.65	95.00	12.83	16.27
c_i^{h}-range	9	0.96	5.65	85.06	9.12	15.30
	999	1.02	7.95	83.83	9.45	16.27
$\sum_{n\in N} m_{i,j}$	$U[0,0.1]$	1.40	7.95	87.08	8.45	12.73
	$U[0,0.5]$	0.59	3.60	81.82	10.12	16.27
All instances		0.99	7.95	84.45	9.28	16.27

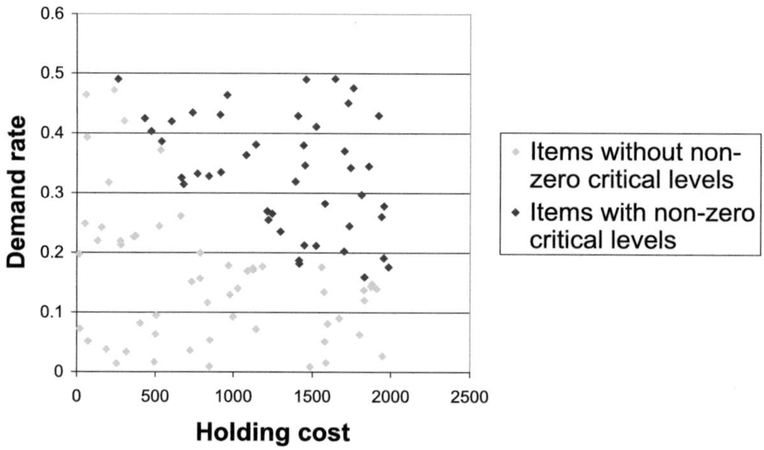

Fig. 4.1 Classification of SKU's in one data set with 2 groups and 100 SKU's

With respect to the savings that are obtained by applying critical level policies, we observe from both tables that average savings slightly increase with the number of SKU's. Furthermore, average savings are a bit larger for the instances with 5 groups. From Table 4.3, we can conclude that larger differences between the target waiting times lead to larger savings, where the instances with a small first group and a large second group apparently lead to the largest savings. Apparently, critical levels are more beneficial in situations where a relatively small part of the customers requires a much lower aggregate mean waiting time than the rest. This observation is interesting since that is what one typically will have in real-life instances. In both tables, it can be seen that larger demand rates lead to an increase in average (and maximum) savings, as do larger values for the average inventory holding cost parameter $\bar{c}^{\mathrm{h}}$. With respect to the larger savings at higher average inventory holding cost parameters: This effect supports the applicability of critical level policies and our method in spare parts inventory management for high-tech equipment, since typically these spare parts are expensive.

The overall average savings is equal to 8–9 %. Maximum savings amount to 31 %. These figures show that substantial savings can be obtained if critical level policies are applied to ration inventory. These relative savings become even more appealing because of the large amount of money invested in spare parts inventory in many industries (see Sect. 1.1). Apparently, it is worthwhile in situations with differentiated target service levels to use this type of policies.

Remark 4.3. For the instances with two machine groups, there is a small subset of 59 instances in which our Dantzig-Wolfe heuristic is slightly more expensive than the obtained basestock policy for the situation with one aggregated customer group (at most 1.15 %). Notice that this is inherent to the use of heuristic solution procedures to compare classes of policies.

4.6 Case Study: ASML

In this section, we present a case study at ASML. We consider the spare parts inventory control for two machine families at one representative local warehouse. We compare the costs obtained under customer differentiation via critical level policies to the costs obtained by simply using basestock policies and providing the lowest target waiting time to all machine groups.

For both machine families, the range of prices (and thus of the inventory holding cost parameters) of the SKU's is large. The ratio between the most expensive and cheapest SKU is in the order of $10^5 - 10^6$ (if the 10 % most expensive SKU's and 10 % cheapest SKU's are excluded, then this ratio is still 10^2).

The first machine family, called Family 1, consists of machines with in total 352 relevant spare parts. We consider 5 machine groups that are completely identical, except for their target aggregate mean waiting times. Those amount to $\frac{1}{3}$, 1, 3, 9, and 27 h, respectively. These are realistic values for the semiconductor supplier industry. The failure rates per SKU for all machine groups together vary from 0.003 to 35.1 per year. The average yearly failure rate is 1.41, and thus very small. The regular replenishment leadtime equals 14 days for all SKU's (i.e., $t_i = 14$, $i \in I$), and emergency shipments take on average 2 days ($t_i^{\text{em}} = 2$, $i \in I$) against additional costs $c_i^{\text{em}} = 1{,}000$ Euros, $i \in I$.

For Family 1, we compare different options in Table 4.5. As a starting point, the round-up policy has been used, i.e., with all 5 customer groups together considered as one group with a target waiting time of $\frac{1}{3}$. The costs of the obtained feasible solution, C_{DW}, has been normalized at 1,000. Besides the situation with 5 customer groups, situations with 2 customer groups have been considered, where in both groups the lowest maximum waiting time of all subgroups is required. These 2 groups can be formed in 4 different sensible ways. In Table 4.5, C_{MP}, C_{DW}, and G are shown for each situation, as well as the savings in comparison to the round-up policy, the number of SKU's with non-zero critical level(s) for the Dantzig-Wolfe heuristic solution, and the computation time.

Table 4.5 Normalized results for Family 1

Number of groups	Composition	C_{MP}	C_{DW}	G (%)	Savings (%)	Number of SKU's with non-zero critical level(s)	Computation time (s)
1		984	1,000	1.64		0	3
2	1 – 2345	946	951	0.59	4.86	21	4
	12 – 345	937	940	0.40	5.96	82	5
	123 – 45	954	956	0.26	4.37	109	5
	1234 – 5	970	977	0.67	2.32	109	5
5		925	926	0.17	7.39	108	10

Table 4.6 Normalized results for Family 2

Number of groups	Composition	C_{MP}	C_{DW}	G (%)	Savings (%)	Number of SKU's with non-zero critical level(s)	Computation time (s)
1		1,000	1,000	0.02		0	5
2	1 – 2345	967	976	0.86	2.45	22	7
	12 – 345	953	958	0.60	4.17	56	8
	123 – 45	965	965	0.03	3.48	134	7
	1234 – 5	983	983	0.00	1.65	135	7
5		937	942	0.48	5.81	125	72

Table 4.6 contains similar results for a second machine family, called Family 2, for which again 5 customer groups are considered, with target average waiting time levels that show less variation: $(\hat{W}_1^{obj}, \hat{W}_2^{obj}, \hat{W}_3^{obj}, \hat{W}_4^{obj}, \hat{W}_5^{obj}) = (0.5, 1, 2, 4, 8)$. In Family 2, we have 465 spare parts with yearly failure rates per SKU for all customer groups together varying from 0.006 to 117, and with an average of 2.28. Again, we have $t_i = 14$ days, $t_i^{em} = 2$ days, and $c_i^{em} = 1{,}000$ Euros, $i \in I$.

The most important observation is that application of our model with customer differentiation may lead to a significant cost reduction in comparison to the round-up policy. As seen in Tables 4.5 and 4.6, the costs for Families 1 and 2 can be reduced by 7.4 % and 5.8 %, respectively, if extensive differentiation is applied, i.e. with a division into 5 customer groups. Taking into account the large amounts of money invested in spare parts inventory in this type of industry – investments are typically in the order of tens of millions of Euros – this constitutes a large potential costs saving in absolute terms. The larger the differentiation in service levels, the larger the potential savings. Notice that in both families, about 75 % of the cost savings of extensive differentiation are already obtained by choosing a smart division into 2 groups. Taking into account the additional implementation efforts and the increased complexity in daily operations if 5 groups are defined, it may be wise in practice to limit the number of customer groups.

As in the earlier computational experiment, the gap G is small in all cases, which implies that we are not far from the optimal solution. Further, computation times are very reasonable. For the instance with 5 groups for Family 2, we plotted the SKU's based on their failure rates and price in Fig. 4.2, and we zoomed in to the lower left area in Fig. 4.3. We observe that SKU's that are cheap and do not fail often generally

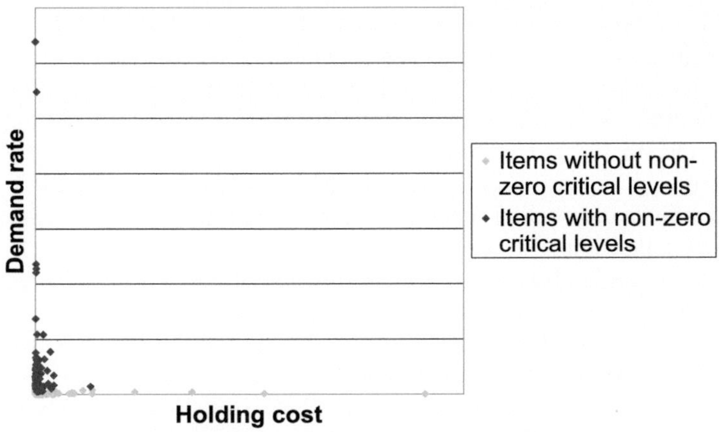

Fig. 4.2 Classification of SKU's in Family 2 data set with 5 groups

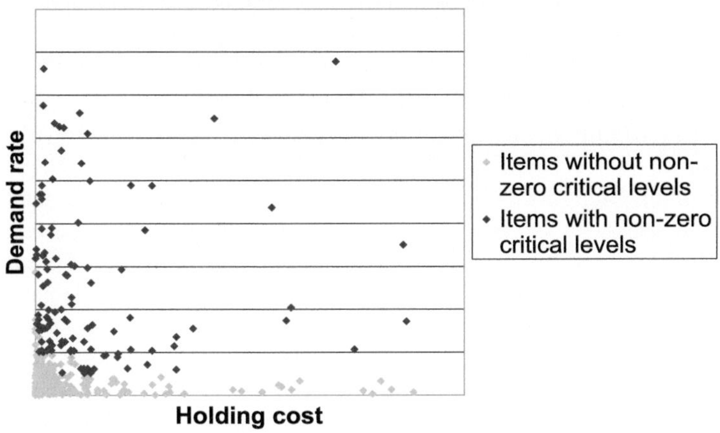

Fig. 4.3 Classification of SKU's in Family 2 data set with 5 groups (detail of lower left area of Fig. 4.2)

have no non-zero critical levels. Studying the critical levels in further detail taught us that where critical levels are non-zero, they usually take values of 1 (only for one SKU in this data set we observed that the critical levels for group 4 and 5 were 2).

4.7 Concluding Remarks

As mentioned in the introduction, in 1965, Veinott [22] was the first study on a problem with multiple demand classes. Since then, many studies on single-item, single-location problems with multiple demand classes have been executed. In one stream of literature, one has derived the optimality of critical level policies, under the assumption of penalty costs for backordered demands or lost sales. For a periodic-review setting with generally distributed demand per period and a zero replenishment leadtime, this was done by Topkis [20]. In a continuous-review setting and under the assumption of a single exponential server that is responsible for the replenishments of the stock, this was done by Ha [12] and De Véricourt et al. [9]. For the case where each replenishment order has its own realisation of the leadtime, drawn from a given distribution function for the replenishment leadtime (this is equivalent to a so-called ample server modeling of the supply), critical level policies are not optimal in general. If however, replenishment leadtimes are exponential and it is assumed that a basestock policy is followed for replenishments, then it is known that using critical levels is optimal for the rationing of the demands; see Miller [17] and Van Jaarsveld [21].

In another stream of literature, authors limit themselves to a class of critical level policies and consider various settings for the optimization within that class, among which the minimization of total costs subject to a service level constraint; see e.g. Dekker et al. [7], Kranenburg and Van Houtum [14], and Möllering and Thonemann [18]. In this stream, there are also papers in which an (s,Q) policy instead of a basestock policy is assumed for the replenishments (see Melchiors et al. [16] and Deshpande et al. [8]), and one has studied a two-class case with lost sales for high-priority demands and backordering for low-priority demands (see Enders et al. [10]).

The work of this chapter is based on [15], and is about the use of critical level policies in a multi-item, single-location setting with a general number of demand classes and system-oriented service level constraints. In this model, we assume that emergency shipments are used for all demands that are not met from stock. So far, only a few other studies are available on similar problems. Alvarez et al. [1] have studied the same problem as in [15], but with a selective use of emergency shipments. This gives an extra dimension to differentiate the service for different demand classes. In Cohen et al. [4, 5], in a periodic-review setting with two demand classes, service differentiation is applied in the sense that in each period the low-priority demand is only fulfilled after the high-priority demand has been fulfilled (no critical levels are being used). They use (s,S)-policies and consider heuristic procedures for the minimization of costs under a system-oriented service level constraint, among which a greedy heuristic.

There is also some work on multiple demand classes in a setting with multiple locations and lateral transshipments; for this stream of work, we refer to the concluding section of the next chapter.

Problems

4.1. Consider the special case in which all average emergency shipment times t_i^{em} are equal. Show that then Problem (P) is equivalent to the same problem with aggregate fill rate constraints instead of aggregate mean waiting time constraints.

4.2. Consider Problem (Q(S)) with

$$
\begin{aligned}
&|N| = 2,\\
&m_1 = 0.25,\ m_2 = 0.75 \ \text{(in demands per month)},\\
&t = 1 \ \text{(in months)},\\
&S = 3,\\
&c_1^{\mathrm{p}} = 3{,}000,\ c_2^{\mathrm{p}} = 200 \ \text{(in Euros)},\\
&c^{\mathrm{h}} = 200 \ \text{(in Euros per part per month)}.
\end{aligned}
$$

The critical level h_1 is equal to 0, and for h_2, it holds that $0 \le h_2 \le S$. Apply Algorithm 4.1 to determine an optimal solution $\mathbf{h}^* = (0, h_2^*)$. What are the corresponding optimal costs?

4.3. Consider Problem (Q(S)) with

$$
\begin{aligned}
&|N| = 2,\\
&m_1 = m_2 = 0.25 \ \text{(in demands per month)},\\
&t = 1 \ \text{(in months)},\\
&S = 2,\\
&c_2^{\mathrm{p}} = \alpha c_1^{\mathrm{p}} = 200 \ \text{(in Euros)},
\end{aligned}
$$

where $0 < \alpha < 1$. The critical level h_1 is equal to 0, and for h_2, it holds that $0 \le h_2 \le S$. Hence, there are three possible solutions for $\mathbf{h}^* = (0, h_2^*)$: (0,0), (0,1), and (0,2).

(a) Determine the average costs for each of the three solutions.
(b) For which values of α is (0,0) optimal, for which values of α is (0,1) optimal, and for which values of α is (0,2) optimal?

4.4. Consider Problem (Q) in Sect. 4.3.3. In this problem we consider an alternative upper bound function for $\tilde{C}(S)$.

(a) Replace all penalty cost parameters c_n^{p} by the largest penalty cost parameter c_1^{p}. Let $\ddot{C}(S)$ be the optimal costs of this adapted problem for each $S \in \mathbb{N}_0$. Explain why this leads to an upper bound for $\tilde{C}(S)$, i.e., why it holds that $\ddot{C}(S) \ge \tilde{C}(S)$.
(b) Derive a cost formula for $\ddot{C}(S)$.
(c) Show that the upper bound $\ddot{C}(S)$ is at most equally good as the upper bound $\tilde{C}^{\mathrm{ub}}(S)$.

4.5. Consider the following generalized version of Problem (P) as described in Sect. 4.2. Instead of an average emergency shipment time t_i^{em} and corresponding cost c_i^{em}, we have emergency shipment times $t_{i,n}^{\mathrm{em}}$ and cost factors $c_{i,n}^{\mathrm{em}}$ that are dependent on the machine group for which an emergency shipment is executed. This may be helpful when one has multiple options for the emergency shipments. Then one may want to apply expensive and fast options for the higher priority machine groups, while cheaper and slower options can be used for lower priority machine groups.

Suppose that one wants to approach the generalized problem in the same way as Problem (P). Denote the changes that have to be made in the formulas and analysis in the Sects. 4.2–4.4.

References

1. Alvarez, E.M., Van der Heijden, M.C., Zijm, W.H.M.: The selective use of emergency shipments for service-contract differentiation. Int. J. Prod. Econ. **143**, 518–526 (2013)
2. Barbour, A.D.: Networks of queues and the method of stages. Adv. Appl. Probab. **8**, 584–591 (1976)
3. Baskett, F., Chandy, K.M., Muntz, R.R., Palacios, F.G.: Open, closed, and mixed networks of queues with different classes of customers. J. Assoc. Comput. Mach. **22**, 248–260 (1975)
4. Cohen, M.A., Kleindorfer, P.R., Lee, H.L.: Near-optimal service constrained stocking policies for spare parts. Oper. Res. **37**, 104–117 (1989)
5. Cohen, M.A., Kleindorfer, P.R., Lee, H.L., Pyke, D.F.: Multi-item service constrained (s,S) policies for spare parts logistics systems. Nav. Res. Logist. **39**, 561–577 (1992)
6. Dantzig, G.B., Wolfe, P.: Decomposition principle for linear programs. Oper. Res. **8**, 101–111 (1960)
7. Dekker, R., Hill, R.M., Kleijn, M.J., Teunter, R.H.: On the $(S-1,S)$ lost sales inventory model with priority demand classes. Nav. Res. Logist. **49**, 593–610 (2002)
8. Deshpande, V., Cohen, M.A., Donohue, K.: A threshold inventory rationing policy for service-differentiated demand classes. Manag. Sci. **49**, 683–703 (2003)
9. De Véricourt, F., Karaesmen, F., Dallery, Y.: Optimal stock allocation for a capacitated supply system. Manag. Sci. **48**, 1486–1501 (2002)
10. Enders, P., Adan, I.J.B.F., Scheller-Wolf, A.A., Van Houtum, G.J.: Inventory rationing for a system with heterogeneous customer classes. Flex. Serv. Manuf. J. **26**, 344–386 (2014)
11. Gnedenko, B.V., Kovalenko, I.N.: Introduction to Queueing Theory. Israel Program for Scientific Translations, Jerusalem (1968)
12. Ha, A.Y.: Inventory rationing in a make-to-stock production system with several demand classes and lost sales. Manag. Sci. **43**, 1093–1103 (1997)
13. Karush, W.: A queueing model for an inventory problem. Oper. Res. **5**, 693–703 (1957)
14. Kranenburg, A.A., Van Houtum, G.J.: Cost optimization in the $(S-1,S)$ lost sales inventory model with multiple demand classes. OR Lett. **35**, 493–502 (2007)
15. Kranenburg, A.A., Van Houtum, G.J.: Service differentiation in spare parts inventory management. J. Oper. Res. Soc. **59**, 946–955 (2008)[1]
16. Melchiors, P., Dekker, R., Kleijn, M.J.: Inventory rationing in an (s,Q) inventory model with lost sales and two demand classes. J. Oper. Res. Soc. **51**, 111–122 (2000)
17. Miller, B.: A queueing reward system with several customer classes. Manag. Sci. **16**, 234–245 (1969)
18. Möllering, K.T., Thonemann, U.W.: An optimal critical level policy for inventory systems with two demand classes. Nav. Res. Logist. **55**, 632–642 (2008)
19. Puterman, M.L.: Markov Decision Processes: Discrete Stochastic Dynamic Programming. Wiley-Interscience, New York (1994)

[1] The material of this paper has been reused in this chapter with kind permission from Palgrave Macmillan.

20. Topkis, D.M.: Optimal ordering and rationing policies in a nonstationary dynamic inventory model with n demand classes. Manag. Sci. **15**, 160–176 (1968)
21. Van Jaarsveld, W.L.: Maintenance centered service parts inventory control. Ph.D. thesis, Erasmus University (2013). http://repub.eur.nl/dissertations
22. Veinott, A.F.: Optimal policy in a dynamic, single product, non-stationary inventory model with several demand classes. Oper. Res. **13**, 761–778 (1965)
23. Wolff, R.W.: Stochastic Modeling and the Theory of Queues. Prentice-Hall International, London (1989)

Chapter 5
Multi-location System with Lateral Transshipments

5.1 Introduction

In this chapter, we consider multiple local warehouses within one geographic region (a part of a continent, a country, or a part of a country). Each local warehouse supports one or multiple groups of installed machines, where each group has its own aggregate mean waiting time target. Machines within one group are identical or close to identical, and they are used by one or multiple customers with the same requirement for the aggregate mean waiting time. Machines in different groups may be (close to) identical, but one may have different aggregate mean waiting time targets for these groups. It is also possible that machines of different groups belong to multiple machine types with common components (without commonality the problem could be decomposed into a problem per machine type). The targets for the aggregate mean waiting time are assumed to be low, which corresponds to high system availabilities, and therefore lateral transshipments and emergency shipments are used in case a machine requires a part at a moment that its local warehouse is out-of-stock.

In practice, one often sees that local warehouses in the same geographic region are replenished from one central warehouse and via consolidated replenishments for all SKU's together. Therefore one has zero or low fixed ordering costs per SKU and one follows basestock control.

For the above situation, one could use the single-location model with emergency shipments of Chap. 3 at a tactical planning level, while still using lateral transshipments at the operational level. This is in fact what several companies do in practice (also commercial software packages for spare parts inventory control generally do *not* include lateral transshipments for tactical planning). However, at the tactical planning level, it is important to plan for *complementary stocks*: if you miss a certain part in your own local warehouse, then you would like to get it from another local warehouse, and vice versa. Therefore, in this chapter, we formulate a multi-item, multi-location, single-echelon model. This model stems from [12] and has

G.J. van Houtum, B. Kranenburg, *Spare Parts Inventory Control under System Availability Constraints*, International Series in Operations Research & Management Science 227, DOI 10.1007/978-1-4899-7609-3_5

been developed in collaboration with ASML. A case study at ASML (carried out in 2004) shows how important it is, in their situation (they have very low demand rates per SKU at local warehouses and therefore many 0–1 values for the basestock levels), to include lateral transshipments at the tactical planning level. This case shows an efficiency improvement of 31 % when the lateral transshipments are included (see Sect. 5.7 for further details). The model of this chapter has been implemented at ASML in 2005 and has led to significantly lower waiting times and lower costs; see also [9]. Later on, the same model has been applied at Océ for the planning of car stocks and so-called quick response stocks; see [20]. Also there an implementation and a significant service and efficiency improvement have been realized.

In our model, we distinguish two types of local warehouses: *main* and *regular* local warehouses. While both main and regular warehouses can receive lateral transshipments, only main local warehouses are allowed to be the supplier of a lateral transshipment. In this way, we get a form of *partial pooling*, with *no pooling* (zero main local warehouses) and *full pooling* (all local warehouses are main local warehouses) as special cases. We will show that most of the benefits of full pooling are obtained with a relatively low number of main local warehouses. This is an important insight for practice. Defining a local warehouse as a main local means that such a warehouse has to have all procedures in place to quickly provide parts for lateral transshipments, which may include a 24/7 availability of manpower. Hence, implementing a system with a low number of main locals is easier and incurs lower fixed costs than a system where all local warehouses are main locals.

While in the models of Chaps. 2–4 a closed-form expression has been obtained for the performance of a given solution, we consider a numerical performance analysis in this chapter. The steady-state behavior can be determined via a Markov process. For sufficiently small instances, one can get the exact performance from the numerically determined steady-state solution of this Markov process. For larger instances, the number of states of the Markov process becomes too large and therefore we also derive an efficient and accurate approximate evaluation procedure. For the determination of the basestock levels, we rely on a greedy heuristic that is similar to the greedy heuristic for the model of Chap. 3.

The organization of this chapter is as follows. In Sect. 5.2, we describe our model. In Sect. 5.3, we explain how a given policy, i.e., a given vector of basestock levels in all local warehouses, can be evaluated exactly. After that, we describe and test our approximate evaluation method in Sect. 5.4. Next, we describe our greedy (heuristic) method to find a feasible policy in Sect. 5.5, followed by a comparison between the performance of partial and full pooling in Sect. 5.6. Then, we present a case study at ASML in Sect. 5.7, and we conclude in Sect. 5.8.

5.2 Model

Let N denote a (non-empty) set of groups of installed machines at customers. Machines in each group $n \in N$ consist of multiple critical components, which we call SKU's. Critical components are subject to failures. Upon a failure, the

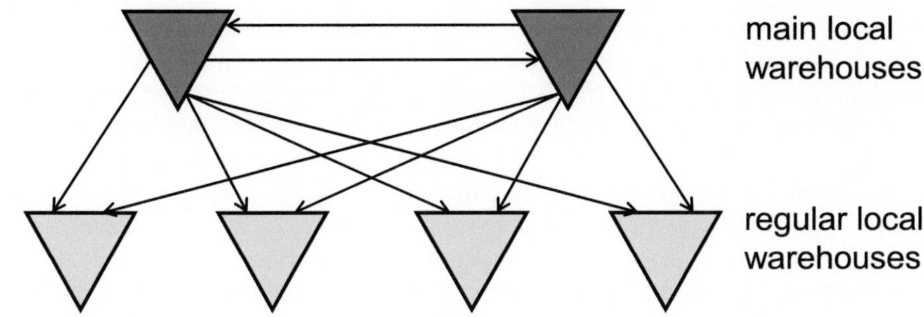

Fig. 5.1 Graphical representation of pooling structure with main and regular local warehouses

corresponding machine becomes non-operational. When the defective part is replaced by a new one, the machine is operational again. Let I denote the (non-empty) set of SKU's; they are numbered as $1,\ldots,|I|$. Demands for each SKU $i \in I$ and group $n \in N$ are assumed to occur according to a Poisson process with a constant rate $m_{i,n}$. We define M_n as the total demand rate for group n, i.e., $M_n := \sum_{i \in I} m_{i,n}$, $n \in N$, and we assume that $M_n > 0$, $n \in N$. The target aggregate mean waiting time for group $n \in N$, i.e., the maximum expected waiting time for an arbitrary request from group n, is denoted by $\hat{W}_n^{\text{obj}}$ (> 0).

Let J denote the (non-empty) set of local warehouses, numbered $j = 1,\ldots,|J|$. We distinguish two types of local warehouses: *main local warehouses* and *regular local warehouses*, shortly referred to as mains and regulars, respectively. Let K $(\subseteq J)$ denote the subset of main local warehouses. All other local warehouses $j \in J \setminus K$ are regular local warehouses. The difference between these two types of local warehouses is that mains can be the supplier of a lateral transshipment whereas regulars cannot. Although in principle K can be empty, in our model description and analysis below we assume $|K| > 0$. If $|K| = 0$, no lateral transshipment takes place, and then the analysis is straightforward because it can be done for each regular local warehouse separately. A graphical representation of the network is given in Fig. 5.1.

Each group $n \in N$ is assigned to exactly one local warehouse $j \in J$ (either a main or a regular local warehouse). This means that if the group needs a spare part because a part of a machine in this group has broken down, it will submit its request to that particular local warehouse. Let N_j $(\subseteq N)$ denote the subset of groups that is assigned to local warehouse $j \in J$; so, the sets N_j, $j \in J$, constitute a partition of N. A local warehouse can serve more than one group; a main local warehouse could have zero groups. Let $M_{i,j}$, $i \in I$, $j \in J$, denote the total demand rate for SKU i at local warehouse j, i.e., $M_{i,j} := \sum_{n \in N_j} m_{i,n}$. If a part is requested by group $n \in N_j$, it will be provided immediately by local warehouse j if this local warehouse has stock on hand. The corresponding waiting time is zero, which also holds for the corresponding cost.

If local warehouse j (either a regular or a main local warehouse) does not have stock on hand, then it tries to obtain the part by means of a lateral transshipment from one of the (other) main local warehouses. The corresponding transportation time for this lateral transshipment from main $k \in K$, $k \neq j$, to local warehouse $j \in J$ is $t_{j,k}^{\text{lat}}$ (≥ 0) and the corresponding cost is $c_{j,k}^{\text{lat}}$ (≥ 0). In case of a need for a lateral transshipment, the main local warehouses are checked in a pre-specified order, and the first main that has stock on hand delivers the part. The pre-specified order has the following structure. Each regular local warehouse is assigned to one main, i.e., in case of a need for a lateral transshipment, this main is checked first. A main can have multiple regulars assigned to it. The main local warehouse $k \in K$ to which a regular local warehouse $j \in J \setminus K$ is assigned, is denoted as k_j. Furthermore, for each main local warehouse $k \in K$, a sequence is given in which all other mains are checked. Thus, the sequence of mains to be checked is identical for demands observed at main $k \in K$ and demands observed at a regular $j \in J \setminus K$ with $k_j = k$. The pre-specified order can be such that you have increasing transportation times, but other orderings are possible as well. Define vector $\sigma(k) := \left(\sigma_1(k), \ldots, \sigma_{|K|-1}(k)\right)$ as the permutation of main local warehouses $K \setminus \{k\}$ that represents this pre-specified order for main local warehouse k (each main local warehouse other than k appears exactly once in this order). Furthermore, let $K(k,\tilde{k})$ ($\subset K$) denote the subset of main local warehouses which are checked earlier than main $\tilde{k}$ according to the pre-specified order for main local warehouse k, i.e., the subset containing all predecessors of $\tilde{k}$ in the pre-specified order for main local warehouse k. Notice that $K(k, \sigma_1(k)) = \emptyset$.

If neither (main or regular) local warehouse j itself nor one of the (other) main local warehouses can deliver the part, an emergency shipment from a central warehouse takes place. The corresponding transportation time and costs (additional costs compared to a normal replenishment) are t_j^{em} ($\geq t_{j,k}^{\text{lat}}, k \in K, k \neq j$), and c_j^{em} ($\geq c_{j,k}^{\text{lat}}, k \in K, k \neq j$), respectively. We assume that the central warehouse has infinite stock of all SKU's and thus can always deliver requested parts. Further, we assume that t_j^{em} and c_j^{em} are significantly larger than the $t_{j,k}^{\text{lat}}$ and $c_{j,k}^{\text{lat}}$, $j \in J$, $k \in K$, $k \neq j$ (think of a factor 1.5 or more for their cost differences). The latter is often the case in practice, and it justifies that: (i) the use of a lateral transshipment is considered before the use of an emergency shipment; (ii) the full stock in a local warehouse is used to provide a lateral transshipment (i.e., it is not needed to use hold back levels, see also Van Wijk et al. [26]). Finally, we assume that the costs and times for lateral transshipments and emergency shipments are mainly determined by the distances, and therefore they are assumed to be the same for all SKU's (i.e., SKU-independent).

The stock in all local warehouses is controlled by a basestock policy. The basestock level for SKU $i \in I$ in local warehouse $j \in J$ is denoted by $S_{i,j}$ ($\in \mathbb{N}_0 = \mathbb{N} \cup \{0\}$). Let $\mathbf{S}_i := (S_{i,1}, \ldots, S_{i,|J|})$, $i \in I$, denote the vector of basestock levels for SKU i, and let a basestock policy for the whole system be denoted by

$$\mathbf{S} = \begin{pmatrix} S_{1,1} & S_{1,2} & \cdots & S_{1,|J|} \\ S_{2,1} & S_{2,2} & \cdots & S_{2,|J|} \\ \vdots & \vdots & \ddots & \vdots \\ S_{|I|,1} & S_{|I|,2} & \cdots & S_{|I|,|J|} \end{pmatrix}$$

Once a part in a local warehouse is used to satisfy a demand, immediately a new part is requested from the central warehouse. This part will be delivered after a certain replenishment leadtime with mean t ($\geq t_j^{\mathrm{em}}, j \in J$). The costs for emergency shipments have been defined as the extra costs in comparison to a normal replenishment. Hence, the costs for the normal replenishments form a constant factor, and therefore they are excluded from our model. The replenishment leadtimes are assumed to be exponentially distributed and the means are the same for all local warehouses. The assumption of equal means is explicitly used in the approximate evaluation method, while this assumption is easily relaxed in the exact evaluation method. It is a justified assumption because in many spare parts networks in practice, replenishment lead times are simply fixed at the same values for all local warehouses in the same geographical area. The assumption of an exponential distribution is made for the following reason. Alfredsson and Verrijdt [1] studied a two-echelon model with lateral transshipment, and they have shown by simulation that their model is to a large extent insensitive to the replenishment lead time distribution for the local warehouses. It is reasonable to assume that this also holds for our model. The assumption of exponential replenishment lead times is explicitly used by the exact evaluation method, while the approximate evaluation method is also applicable for generally distributed replenishment lead times.

The cost for holding one part of SKU $i \in I$ in stock for one time unit is c_i^{h}. We assume that inventory holding costs are also incurred for parts in replenishment. This is a logical choice when the whole spare parts supply chain is owned by one company. It is relatively easy to change this assumption and to exclude the parts in replenishment for the inventory holding costs (see Problem 5.4).

With respect to the fulfillment of a demand for SKU $i \in I$ at local warehouse $j \in J$, we introduce the following notation:

- $\beta_{i,j}(\mathbf{S}_i)$, for the fraction of demand for SKU i at local warehouse j that is delivered immediately upon request, i.e., from the stock in local warehouse j itself, also called the *(item) fill rate*;
- $\alpha_{i,j,k}(\mathbf{S}_i)$, $k \in K$, $k \neq j$, for the fraction of demand for SKU i at local warehouse j that is delivered from main local warehouse k by means of a lateral transshipment;
- $\theta_{i,j}(\mathbf{S}_i)$, for the fraction of demand for SKU i at local warehouse j that is delivered from the central warehouse by an emergency shipment.

Besides we define

$$A_{i,j}(\mathbf{S}_i) := \sum_{k \in K, k \neq j} \alpha_{i,j,k}(\mathbf{S}_i)$$

as the total fraction of demand for SKU $i \in I$ at local warehouse $j \in J$ that is delivered by a lateral transshipment. Notice that for each SKU $i \in I$ at each local warehouse $j \in J$, it holds that

$$\beta_{i,j}(\mathbf{S}_i) + A_{i,j}(\mathbf{S}_i) + \theta_{i,j}(\mathbf{S}_i) = 1. \tag{5.1}$$

Furthermore, notice that $\theta_{i,k}(\mathbf{S}_i)$ is equal for all mains $k \in K$ because an emergency transshipment from the central warehouse only takes place if none of the main local warehouses has stock on hand, and, obviously, this probability is independent on which main local warehouse observes the demand.

In Sects. 5.3 and 5.4, we show how all fractions can be determined, in an exact and approximate way, respectively, for a given basestock policy $\mathbf{S}_i$ for SKU $i \in I$.

Let $W_{i,j}(\mathbf{S}_i)$, $i \in I$, $j \in J$, denote the mean waiting time if SKU i is requested at local warehouse j, under a given basestock policy $\mathbf{S}_i$ for SKU i. Then, $W_{i,j}(\mathbf{S}_i)$ can be calculated as follows:

$$W_{i,j}(\mathbf{S}_i) = \sum_{k \in K, k \neq j} t^{\text{lat}}_{j,k} \alpha_{i,j,k}(\mathbf{S}_i) + t^{\text{em}}_j \theta_{i,j}(\mathbf{S}_i), \tag{5.2}$$

For a group of machines n connected to local warehouse $j \in J$ (i.e., $n \in N_j$), the aggregate mean waiting time equals

$$\hat{W}_n(\mathbf{S}) = \sum_{i \in I} \frac{m_{i,n}}{M_n} W_{i,j}(\mathbf{S}_i).$$

For SKU $i \in I$, the total inventory holding costs per time unit are equal to

$$\sum_{j \in J} c^{\text{h}}_i S_{i,j}, \tag{5.3}$$

and the expected total transportation costs for lateral transshipments and emergency shipments are equal to

$$\sum_{j \in J} M_{i,j} \left(\sum_{k \in K, k \neq j} c^{\text{lat}}_{j,k} \alpha_{i,j,k}(\mathbf{S}_i) + c^{\text{em}}_j \theta_{i,j}(\mathbf{S}_i) \right).$$

Therefore, for SKU $i \in I$, the expected total cost per time unit, $C_i(\mathbf{S}_i)$, is

$$C_i(\mathbf{S}_i) = \sum_{j \in J} c^{\text{h}}_i S_{i,j} + \sum_{j \in J} M_{i,j} \left(\sum_{k \in K, k \neq j} c^{\text{lat}}_{j,k} \alpha_{i,j,k}(\mathbf{S}_i) + c^{\text{em}}_j \theta_{i,j}(\mathbf{S}_i) \right), \tag{5.4}$$

and the total average costs are equal to $C(\mathbf{S}) = \sum_{i \in I} C_i(\mathbf{S}_i)$.

The objective is to minimize the expected total cost for all SKU's together, under the condition that the expected waiting time for an arbitrary request from each group $n \in N$ does not exceed the target aggregate mean waiting time $\hat{W}^{\text{obj}}_n$. The problem can be mathematically formulated as follows:

$$\text{(P)} \qquad \min \qquad C(\mathbf{S})$$

$$\text{subject to } \hat{W}_n(\mathbf{S}) \leq \hat{W}^{\text{obj}}_n, \quad j \in J, n \in N_j,$$

$$\mathbf{S} \in \mathscr{S},$$

where $\mathscr{S} = \{\mathbf{S} = (S_{i,j})_{i \in I, j \in J} \mid S_{i,j} \in \mathbb{N}_0, \forall i \in I \text{ and } j \in J\}$.

In the next sections, Sects. 5.3–5.5, we describe how a given policy can be evaluated, either exactly or approximately, and how a feasible solution for Problem (P) can be obtained.

5.3 Exact Evaluation

In this section, we describe how a given basestock policy can be evaluated exactly. Evaluation can be done for each SKU $i \in I$ separately. In the evaluation, for a given policy $\mathbf{S}_i$ for SKU i, i.e., for a given vector of basestock levels in all local warehouses $j \in J$ for SKU i, we determine $\beta_{i,j}(\mathbf{S}_i)$, $j \in J$, $\alpha_{i,j,k}(\mathbf{S}_i)$, $j \in J$, $k \in K$, $k \neq j$, and $\theta_{i,j}(\mathbf{S}_i)$, $j \in J$. Once we have determined those, the mean waiting times $W_{i,j}(\mathbf{S}_i)$, $j \in J$, and costs $C_i(\mathbf{S}_i)$, follow immediately from Eqs. (5.2) and (5.4).

For the exact evaluation of a given policy for SKU $i \in I$, we use a Markov process description. The states may be described by the on-hand stocks at all the warehouses $j \in J$. This gives a $|J|$-dimensional state space, and thus exact numerical evaluation can only be done for a limited number of local warehouses, because the number of states, and therefore also the computation time, grows exponentially in the number of local warehouses.

For our Markov process, let $\mathbf{x}_i = (x_{i,1}, \ldots, x_{i,|J|})$ denote a vector that represents the current state, where $x_{i,j}$, $j \in J$, is the number of parts on hand in local warehouse j. Obviously, $0 \leq x_{i,j} \leq S_{i,j}$, $j \in J$. Furthermore, let $\mathbf{e}_j$ denote a row vector of size $|J|$ with the j-th element equal to 1 and all other elements equal to 0.

If the system is in state $\mathbf{x}_i$, then two types of events can occur. First, at any local warehouse $j \in J$, a request can arrive for SKU i. Secondly, at each local warehouse $j \in J$ that has less stock on hand than its basestock level ($x_{i,j} < S_{i,j}$), a part of SKU i can arrive as a stock replenishment from the central warehouse. By considering both types of events for all applicable local warehouses, we obtain the transition rates from state x_i to other states. To identify all transitions, this has to be done for all states.

If a part of SKU i is requested at a regular local warehouse $j \in J \setminus K$ while the system is in state $\mathbf{x}_i$, then one of the following cases applies:

- Regular local warehouse j has stock on hand. In this case, the regular local warehouse can supply the requested part from its own stock. The corresponding transition is $\mathbf{x}_i \rightarrow \mathbf{x}_i - \mathbf{e}_j$, and the transition rate is $M_{i,j}$.
- Regular local warehouse j has no stock on hand, but main local warehouse k_j has stock on hand. In this case, main local warehouse k_j delivers the requested part. The corresponding transition is $\mathbf{x}_i \rightarrow \mathbf{x}_i - \mathbf{e}_{k_j}$, and the transition rate is $M_{i,j}$.
- Neither regular local warehouse j nor main local warehouse k_j has stock on hand, but at least one other main local warehouse has stock on hand. In this case, the part is delivered by the first main local warehouse k in the predefined lateral transshipment order $\sigma(k_j)$ that has stock on hand. The corresponding transition is $\mathbf{x}_i \rightarrow \mathbf{x}_i - \mathbf{e}_k$, and the transition rate is $M_{i,j}$.

- Neither regular local warehouse j nor any of the main local warehouses has stock on hand. In this case, the part will be supplied by an emergency replenishment from the central warehouse, and the system remains in the current state (no transition takes place).

If a part of SKU i is requested at a main local warehouse $j \in K$ while the system is in state $\mathbf{x}_i$, then one of the following cases applies:

- Main local warehouse j has stock on hand. In this case, the main local warehouse can supply the requested part from its own stock. The corresponding transition is $\mathbf{x}_i \to \mathbf{x}_i - \mathbf{e}_j$, and the transition rate is $M_{i,j}$.
- Main local warehouse j has no stock on hand, but at least one other main local warehouse has stock on hand. In this case, the part is delivered by the first main local warehouse k in the predefined lateral transshipment order $\sigma(j)$ that has stock on hand. The corresponding transition is $\mathbf{x}_i \to \mathbf{x}_i - \mathbf{e}_k$, and the transition rate is $M_{i,j}$.
- None of the main local warehouses has stock on hand. In this case, the part will be supplied by an emergency replenishment from the central warehouse, and the system remains in the current state (no transition takes place).

If a part of SKU i arrives as stock replenishment in local warehouse $j \in J$ with $x_{i,j} < S_{i,j}$ while the system is in state $\mathbf{x}_i$, then the transition is $\mathbf{x}_i \to \mathbf{x}_i + \mathbf{e}_j$, and the transition rate is $(S_{i,j} - x_{i,j})/t$. The numerator is the number of parts of local warehouse j that is currently being replenished.

When all transitions have been determined, we can apply uniformization (see e.g. [21], Sect. 5.8) by adding transitions from a state to itself. After uniformization, we can find the steady-state distribution, i.e., all steady-state probabilities, for our Markov process by successive approximation, also known as the power method (see e.g. [29], p. 281). Theoretically, successive approximation provides us with exact steady-state probabilities. Numerically, however, we will stop when steady-state probabilities do not change more than ε, with ε small (e.g., $\varepsilon = 10^{-6}$). For all states $\mathbf{x}_i$, let $\pi(\mathbf{x}_i)$ denote the steady-state probability that the Markov process is in state $\mathbf{x}_i$.

The next step is to determine $\beta_{i,j}(\mathbf{S}_i)$, $\alpha_{i,j,k}(\mathbf{S}_i)$, $k \in K$, $k \neq j$, and $\theta_{i,j}(\mathbf{S}_i)$, for all local warehouses $j \in J$. Remember that for each local warehouse j, these values sum up to 1. Each steady-state probability $\pi(\mathbf{x}_i)$ contributes to exactly one of the terms $\beta_{i,j}(\mathbf{S}_i)$, $\alpha_{i,j,k}(\mathbf{S}_i)$, and $\theta_{i,j}(\mathbf{S}_i)$, and thus they can be determined by summation of the correct probabilities. For each regular local warehouse $j \in J \setminus K$, it holds that

$$\beta_{i,j}(\mathbf{S}_i) = \sum_{\mathbf{x}_i | x_{i,j}>0} \pi(\mathbf{x}_i);$$

$$\alpha_{i,j,k_j}(\mathbf{S}_i) = \sum_{\mathbf{x}_i | x_{i,j}=0,\ x_{i,k_j}>0} \pi(\mathbf{x}_i);$$

$$\alpha_{i,j,k}(\mathbf{S}_i) = \sum_{\mathbf{x}_i | x_{i,j}=0,\ x_{i,k_j}=0,\ x_{i,\hat{k}}=0\ (\forall \hat{k} \in K(k_j,k)),\ x_{i,k}>0} \pi(\mathbf{x}_i), \quad k \in K, k \neq k_j;$$

$$\theta_{i,j}(\mathbf{S}_i) = \sum_{\mathbf{x}_i | x_{i,j}=0,\ x_{i,k}=0\ (\forall k \in K)} \pi(\mathbf{x}_i).$$

For each main local warehouse $j \in K$, we obtain:

$$\beta_{i,j}(\mathbf{S}_i) = \Sigma_{\mathbf{x}_i | x_{i,j}>0}\ \pi(\mathbf{x}_i);$$

$$\alpha_{i,j,k}(\mathbf{S}_i) = \Sigma_{\mathbf{x}_i | x_{i,j}=0,\ x_{i,\hat{k}}=0\ (\forall \hat{k} \in K(j,k)),\ x_{i,k}>0}\ \pi(\mathbf{x}_i), \quad k \in K, \quad k \neq j;$$

$$\theta_{i,j}(\mathbf{S}_i) = \Sigma_{\mathbf{x}_i | x_{i,k}=0\ (\forall k \in K)}\ \pi(\mathbf{x}_i).$$

Of course, for each local warehouse $j \in J$, one of the terms $\beta_{i,j}(\mathbf{S}_i)$, $\alpha_{i,j,k}(\mathbf{S}_i)$, and $\theta_{i,j}(\mathbf{S}_i)$, could be calculated alternatively as the difference between 1 and the sum of the other terms.

5.4 Approximate Evaluation

In Sect. 5.3, we have described how a basestock policy can be evaluated exactly. However, exact evaluation is done numerically, and can be time-consuming or even computationally intractable if the number of local warehouses is large, since each local warehouse constitutes a dimension in the Markov process. (Of course, in the special case that all local warehouses are regulars, i.e., if no lateral transshipment takes place at all, each regular can be analyzed individually and we do not have computational problems.) Since we want our method to be applicable for real-life instances with many local warehouses, we introduce an approximate evaluation method in this section.

As mentioned in the previous section, evaluation can be done for each SKU $i \in I$ separately. In the approximate evaluation, for a given policy $\mathbf{S}_i$ for SKU i, we approximate the $\beta_{i,j}(\mathbf{S}_i)$, $\alpha_{i,j,k}(\mathbf{S}_i)$, and $\theta_{i,j}(\mathbf{S}_i)$; the mean waiting times $W_{i,j}(\mathbf{S}_i)$ and costs $C_i(\mathbf{S}_i)$ again follow from Eqs. (5.2) and (5.4).

In the approximate evaluation, we make use of the loss probability in the Erlang loss model (i.e., the $M|G|c|c$ queue). Let $L(c,\rho)$ denote this loss probability, where $c \in \mathbb{N}_0$ represents the number of servers and $\rho > 0$ the offered load. Then $L(c,\rho)$ is given by (see e.g. [25])

$$L(c,\rho) = \frac{\rho^c/c!}{\Sigma_{x=0}^{c} \rho^x/x!}$$

(by convention, $L(0,\rho) = 1$, $\rho > 0$).

The key to our approximate evaluation method is that: (i) We decouple the network into individual local warehouses; (ii) We assume that the extra demand processes in main local warehouses due to requests for lateral transshipment are Poisson processes. The decoupling is done in two steps. In the first step, we decouple the regular local warehouses from the mains. In the second step the resulting system of main local warehouses is decoupled, so that each main can be analyzed individually. These two decoupling steps are discussed in Sects. 5.4.1 and 5.4.2, respectively. After that, we describe our approximate evaluation method more formally in Sect. 5.4.3. In Sect. 5.4.4, we compare our approximate evaluation to exact evaluation.

5.4.1 Decoupling the Regulars from the Mains

The connection between regular local warehouses $j \in J \setminus K$ and all main local warehouses $k \in K$ is that the mains can provide a lateral transshipment to a regular. A demand for such a lateral transshipment occurs when a regular j faces a demand at a moment that it is out of stock. We refer to the demand process for lateral transshipments as generated by regular $j \in J \setminus K$ as an *overflow demand process*. In our approximate evaluation, *we assume that the overflow demand process at regular $j \in J \setminus K$ behaves as a Poisson process that constitutes an additional demand stream at main k_j*. Obviously, in reality this demand process can be burstier than Poisson, as it occurs only at the moments that the regular local warehouse is out of stock, but for low demand rates our assumption seems reasonable. Also, notice that in case a regular j has a zero basestock level, i.e., $S_{i,j} = 0$, the overflow demand really follows a Poisson process, since all demand is forwarded.

First, each regular $j \in J \setminus K$ is analyzed separately using the Erlang loss model. The steady-state behavior of the on-hand stock in a regular local warehouse j is identical to the steady-state behavior of the number of idle servers in an Erlang loss system with $S_{i,j}$ servers (i.e., a part on stock is as a free server and an outstanding replenishment order is as an occupied server), arrival rate $M_{i,j}$, and mean service time t. Hence, $\beta_{i,j}(\mathbf{S}_i)$ (which actually only depends on $S_{i,j}$, and not on the basestock levels at other local warehouses) is equal to the corresponding Erlang loss probability, i.e., $\beta_{i,j}(\mathbf{S}_i) = 1 - L(S_{i,j}, M_{i,j}t)$ (so, this formula is exact).

At this point, we use our assumption that the demand for lateral transshipments placed by regular j at main k_j follows a Poisson process; the corresponding rate is $(1 - \beta_{i,j}(\mathbf{S}_i))M_{i,j}$. We introduce $\tilde{M}_{i,k}$ for the demand rate for SKU i at main local warehouse $k \in K$ *including* all demand for lateral transshipment from regulars assigned to this main, i.e.,

$$\tilde{M}_{i,k} := M_{i,k} + \sum_{j \in J | k_j = k} (1 - \beta_{i,j}(\mathbf{S}_i)) M_{i,j}.$$

We can now analyze the system of mains, where each main $k \in K$ faces a Poisson demand process with parameter $\tilde{M}_{i,k}$; this analysis is described in Sect. 5.4.2. The analysis of the system of mains provides us with $\beta_{i,k}(\mathbf{S}_i)$, $k \in K$, $\alpha_{i,k,\tilde{k}}(\mathbf{S}_i)$, $k \in K$, $\tilde{k} \in K$, $\tilde{k} \neq k$, and $\theta_{i,k}(\mathbf{S}_i)$, $k \in K$.

After the analysis of the system of mains, we can determine the remaining performance measures $\alpha_{i,j,k}(\mathbf{S}_i)$, $k \in K$, and $\theta_{i,j}(\mathbf{S}_i)$ for all regulars $j \in J \setminus K$:

$$\alpha_{i,j,k}(\mathbf{S}_i) := \begin{cases} (1 - \beta_{i,j}(\mathbf{S}_i))\beta_{i,k_j}(\mathbf{S}_i), & k = k_j, \\ (1 - \beta_{i,j}(\mathbf{S}_i))\alpha_{i,k_j,k}(\mathbf{S}_i), & k \in K, k \neq k_j, \end{cases} \tag{5.5}$$

and $\theta_{i,j}(\mathbf{S}_i) := (1 - \beta_{i,j}(\mathbf{S}_i))\theta_{i,k_j}(S_i)$.

Notice that, under the above assumption for the overflow demand processes, the system of mains could be analyzed exactly, by the method described in Sect. 5.3, in which case the computation time is determined by the numerical solution of the steady-state equations of a $|K|$-dimensional Markov process. This could be an attractive alternative for sufficiently low values of $|K|$ (≤ 3, say) and the basestock levels $S_{i,k}$, $k \in K$. Generally, we will follow a further decomposition, which we describe next.

5.4.2 Decoupling the Mains

In the second decoupling step, we decouple the main local warehouses, so that each main can be analyzed separately. The connection between the main local warehouses $k \in K$ is that lateral transshipments can take place from each main to each other main. A demand for a lateral transshipment occurs when a main $k \in K$ faces customer demand at a moment that it is out of stock. We refer to the demand process for lateral transshipments arising at main $k \in K$ as the *overflow demand process* at main $k \in K$. In our approximate evaluation, *we assume that the overflow demand process at main $k \in K$ behaves as a Poisson process that constitutes additional demand streams at all other mains l, $l \in K$, $l \neq k$.* Again, in reality this demand process could be burstier than Poisson, as it occurs only at the moments that main k is out of stock. However, by making the assumption that it follows a Poisson process, we can decouple the mains, and analyze each main individually. For the analysis of each main local warehouse, we use the Erlang loss model.

Recall that the demand processes at the main local warehouses $k \in K$ are approximated by Poisson processes with rates $\tilde{M}_{i,k}$ (where the overflow processes from the regular local warehouses are included) and they have basestock levels $S_{i,k}$. First, under the assumption that these demand processes are exactly equal to Poisson processes (we have this in special cases, e.g. when we have a system with mains only), we can determine the $\theta_{i,k}(\mathbf{S}_i)$ exactly by considering the *aggregate* system of all main local warehouses $k \in K$ together. Then, because of full pooling of inventory between the main local warehouses and the application of emergency shipments in case all main local warehouses are out of stock, the steady-state behavior of the total stock in the aggregate system is identical to the steady-state behavior of the number of idle servers in an Erlang loss system with $\sum_{k \in K} S_{i,k}$ servers, arrival rate $\sum_{k \in K} \tilde{M}_{i,k}$, and mean service time t. Thus, we can calculate $\theta_{i,k}(\mathbf{S}_i)$, $k \in K$, as the Erlang loss probability of the aggregate system:

$$\theta_{i,k}(\mathbf{S}_i) := L\left(\sum_{k' \in K} S_{i,k'}, \sum_{k' \in K} \tilde{M}_{i,k'} t\right), \quad k \in K.$$

So, in special cases, this expression is exact. In general, the demand processes at the main local warehouses $k \in K$ will be almost equal to Poisson processes, but not exactly, and then this expression is an approximation.

At this point, we introduce $\hat{M}_{i,\tilde{k},k}$, $k, \tilde{k} \in K$, $\tilde{k} \neq k$, for the rate with which main $\tilde{k}$ requests a lateral transshipment from main k. Furthermore, we introduce $\hat{M}_{i,k}$ for the demand rate for SKU i at main local warehouse $k \in K$ including *all* demand for lateral transshipments. Thus,

$$\hat{M}_{i,k} := \tilde{M}_{i,k} + \sum_{\tilde{k}\in K, \tilde{k}\neq k} \hat{M}_{i,\tilde{k},k}, \qquad k \in K.$$

In our approximate evaluation, we analyze each main local warehouse separately, using the Erlang loss model. For each main $k \in K$,

$$\begin{aligned} \beta_{i,k}(\mathbf{S}_i) &:= 1 - L\left(S_{i,k}, \hat{M}_{i,k}t\right), \\ A_{i,k}(\mathbf{S}_i) &:= 1 - \left(\beta_{i,k}(\mathbf{S}_i) + \theta_{i,k}(\mathbf{S}_i)\right). \end{aligned}$$

So, $\beta_{i,k}(\mathbf{S}_i)$ and $A_{i,k}(\mathbf{S}_i)$ are dependent on $\hat{M}_{i,k}$. (It is discussed later how we obtain the values for $\alpha_{i,k,\tilde{k}}(\mathbf{S}_i)$, $\tilde{k} \in K, \tilde{k} \neq k$.)

On the other hand, $\hat{M}_{i,k}$, $k \in K$, is dependent on fill rates. We determine its components $\hat{M}_{i,\tilde{k},k}$, $\tilde{k} \in K, \tilde{k} \neq k$, as:

$$\hat{M}_{i,\tilde{k},k} := \begin{cases} \dfrac{A_{i,\tilde{k}}(\mathbf{S}_i)\tilde{M}_{i,\tilde{k}}}{1 - \prod_{\ell\in K,\ell\neq\tilde{k}}\left(1 - \beta_{i,\ell}(\mathbf{S}_i)\right)} \prod_{\ell\in K(\tilde{k},k)} \left(1 - \beta_{i,\ell}(\mathbf{S}_i)\right), \\ \qquad\qquad S_{i,\ell} > 0 \text{ for at least one } \ell \in K\setminus\{\tilde{k}\}, \\ 0, \qquad\qquad \text{otherwise,} \end{cases} \tag{5.6}$$

with the product term $\prod_{\ell\in K(\tilde{k},k)}\left(1 - \beta_{i,\ell}(\mathbf{S}_i)\right)$ defined as 1 if $K(\tilde{k},k) = \emptyset$. The explanation for this formula follows below.

We explain Eq. (5.6) by means of an example. Consider a system with 3 mains, numbered 1, 2, and 3, with $\sigma(1) = (2,3)$. We derive Eq. (5.6) for $\hat{M}_{i,1,2}$ and $\hat{M}_{i,1,3}$ for the case that $S_{i,2} > 0$ or $S_{i,3} > 0$. In the Erlang loss model for main 2, a fraction $\beta_{i,2}(\mathbf{S}_i)$ of the observed Poisson demand rate $\hat{M}_{i,1,2}$ for lateral transshipments as requested by main 1 will be fulfilled. According to the pre-defined lateral transshipment order $\sigma(1)$, we let main 3 face the remaining demand for lateral transshipments to main 1, i.e., $\hat{M}_{i,1,3} := (1 - \beta_{i,2}(\mathbf{S}_i))\hat{M}_{i,1,2}$, again assuming that this demand follows a Poisson process. In the Erlang loss model for main 3, a fraction $\beta_{i,3}(\mathbf{S}_i)$ of the observed Poisson demand rate $\hat{M}_{i,1,3}$ for lateral transshipments as requested by main 1 will be fulfilled. Under the assumption of independency of the physical stocks in mains 2 and 3, we have that (in expectation) in total the following amount is satisfied by means of a lateral transshipment:

$$\begin{aligned} &\beta_{i,2}(\mathbf{S}_i)\hat{M}_{i,1,2} + \beta_{i,3}(\mathbf{S}_i)\hat{M}_{i,1,3} \\ &= \beta_{i,2}(\mathbf{S}_i)\hat{M}_{i,1,2} + \beta_{i,3}(\mathbf{S}_i)(1 - \beta_{i,2}(\mathbf{S}_i))\hat{M}_{i,1,2} \\ &= \left(\beta_{i,2}(\mathbf{S}_i) + \beta_{i,3}(\mathbf{S}_i) - \beta_{i,2}(\mathbf{S}_i)\beta_{i,3}(\mathbf{S}_i)\right)\hat{M}_{i,1,2} \\ &= \left[1 - (1 - \beta_{i,2}(\mathbf{S}_i))(1 - \beta_{i,3}(\mathbf{S}_i))\right]\hat{M}_{i,1,2}. \end{aligned} \tag{5.7}$$

Because in total a fraction $A_{i,1}(\mathbf{S}_i)$ of the demand rate $\tilde{M}_{i,1}$ at main 1 is supplied by a lateral transshipment from the other mains 2 and 3, the expression in (5.7) equals $A_{i,1}(\mathbf{S}_i)\tilde{M}_{i,1}$, or, equivalently,

$$\hat{M}_{i,1,2} = \frac{A_{i,1}(\mathbf{S}_i)\tilde{M}_{i,1}}{1-(1-\beta_{i,2}(\mathbf{S}_i))\,(1-\beta_{i,3}(\mathbf{S}_i))}. \tag{5.8}$$

And thus,

$$\hat{M}_{i,1,3} = \frac{A_{i,1}(\mathbf{S}_i)\tilde{M}_{i,1}}{1-(1-\beta_{i,2}(\mathbf{S}_i))\,(1-\beta_{i,3}(\mathbf{S}_i))}(1-\beta_{i,2}(\mathbf{S}_i)). \tag{5.9}$$

Equations (5.8) and (5.9) are in accordance with Eq. (5.6).

If $S_{i,\ell} = 0$ for all $\ell \in K \setminus \tilde{k}$, then $\hat{M}_{i,\tilde{k},k} = 0$. We distinguish this case since then the denominator $1 - \prod_{\ell \in K, \ell \neq \tilde{k}} (1 - \beta_{i,\ell}(\mathbf{S}_i))$ in Eq. (5.6) would be zero. Notice that in this case, no lateral transshipment to main $\tilde{k}$ will take place at all.

We use an *iterative procedure* to determine $\hat{M}_{i,k}$, $k \in K$, $\hat{M}_{i,\tilde{k},k}$, $k, \tilde{k} \in K$, $\tilde{k} \neq k$, $\beta_{i,k}(\mathbf{S}_i)$, $k \in K$, and $A_{i,k}(\mathbf{S}_i)$, $k \in K$. This procedure works as follows. Initially, we suppose that no lateral transshipment takes place between the mains and accordingly, we set the demand rates equal to $\hat{M}_{i,k} := \tilde{M}_{i,k}$, $k \in K$. We apply the Erlang loss model for each main $k \in K$ and determine $\beta_{i,k}(\mathbf{S}_i) := 1 - L\left(S_{i,k}, \hat{M}_{i,k}t\right)$, and $A_{i,k}(\mathbf{S}_i) := 1 - \left(\beta_{i,k}(\mathbf{S}_i) + \theta_{i,k}(\mathbf{S}_i)\right)$, $k \in K$. For *one* main k', we now determine $\hat{M}_{i,\tilde{k},k'}$, $\tilde{k} \in K, \tilde{k} \neq k'$, using Eq. (5.6), and, subsequently, $\hat{M}_{i,k'}$, $\beta_{i,k'}(\mathbf{S}_i)$, and $A_{i,k'}(\mathbf{S}_i)$. We repeat this for all other mains $k \in K$, and then consider main k' again, etcetera. We continue this iterative procedure until each of the $\hat{M}_{i,k}$, $k \in K$, does not change more than ε, where ε is a small positive number ($\varepsilon = 10^{-6}$, say). We observed that this iterative procedure converges in all cases, but there is no formal proof for the convergence. After convergence, the values for $\alpha_{i,k,\tilde{k}}(S_i)$, $k \in K$, $\tilde{k} \in K$, $\tilde{k} \neq k$, are determined as

$$\alpha_{i,k,\tilde{k}}(\mathbf{S}_i) := \beta_{i,\tilde{k}}(\mathbf{S}_i)\hat{M}_{i,k,\tilde{k}}/\tilde{M}_{i,k}.$$

5.4.3 Formal Description of the Approximate Evaluation Method

In Sects. 5.4.1 and 5.4.2, we have described two decoupling steps that we use in our approximate evaluation method. In the current section, we describe the whole method formally. We describe two algorithms, Algorithms 5.1 and 5.2. Algorithm 5.2 describes Step 3 of Algorithm 5.1 in detail.

Algorithm 5.1

Step 1 For all regulars $j \in J \setminus K$, $\beta_{i,j}(\mathbf{S}_i) := 1 - L(S_{i,j}, M_{i,j}t)$.

Step 2 For all mains $k \in K$, $\tilde{M}_{i,k} := M_{i,k} + \sum_{j \in J | k_j = k} (1 - \beta_{i,j}(\mathbf{S}_i))\, M_{i,j}$.

Step 3 For all mains $k \in K$, determine $\beta_{i,k}(\mathbf{S}_i)$, $\alpha_{i,k,\tilde{k}}(\mathbf{S}_i)$, $\tilde{k} \in K, \tilde{k} \neq k$, and $\theta_{i,k}(\mathbf{S}_i)$, using Algorithm 5.2.

Step 4 For all regulars $j \in J \setminus K$, if $K = \emptyset$, then $\theta_{i,j}(\mathbf{S}_i) := (1 - \beta_{i,j}(\mathbf{S}_i))$. Otherwise, $\alpha_{i,j,k}(\mathbf{S}_i)$ is determined using Eq. (5.5) and $\theta_{i,j}(\mathbf{S}_i) := (1 - \beta_{i,j}(\mathbf{S}_i))\theta_{i,k_j}(\mathbf{S}_i)$.

Algorithm 5.2

Step 1 For all mains $k \in K$, $\theta_{i,k}(\mathbf{S}_i) := L\left(\sum_{k \in K} S_{i,k}, \sum_{k \in K} \tilde{M}_{i,k} t\right)$.
Step 2 For all mains $k \in K$, $\beta_{i,k}(\mathbf{S}_i) := 1 - L\left(S_{i,k}, \tilde{M}_{i,k} t\right)$, and
$A_{i,k}(\mathbf{S}_i) := 1 - \left(\beta_{i,k}(\mathbf{S}_i) + \theta_{i,k}(\mathbf{S}_i)\right)$.
Step 3 For *one main* $k \in K$:

Step 3-a Determine $\hat{M}_{i,\tilde{k},k}$ using Eq. (5.6), and $\hat{M}_{i,k} := \tilde{M}_{i,k} + \sum_{\tilde{k} \in K, \tilde{k} \neq k} \hat{M}_{i,\tilde{k},k}$.
Step 3-b $\beta_{i,k}(\mathbf{S}_i) := 1 - L\left(S_{i,k}, \hat{M}_{i,k} t\right)$, and $A_{i,k}(\mathbf{S}_i) := 1 - \left(\beta_{i,k}(\mathbf{S}_i) + \theta_{i,k}(\mathbf{S}_i)\right)$.

Step 4 Repeat Step 3 for all other mains $k \in K$.
Step 5 Repeat Steps 3 and 4 until $\hat{M}_{i,k}$ does not change more than ε for each $k \in K$, with ε small.
Step 6 For all mains $k \in K$, $\alpha_{i,k,\tilde{k}}(\mathbf{S}_i) := \beta_{i,\tilde{k}}(\mathbf{S}_i)\hat{M}_{i,k,\tilde{k}}/\hat{M}_{i,k}$, $\tilde{k} \in K$, $\tilde{k} \neq k$.

Notice that in the full pooling situation, i.e., if all local warehouses are mains, our algorithm basically boils down to Algorithm 5.2. In the no pooling situation, i.e., if all local warehouses are regulars, then each regular is analyzed individually (using only Steps 1 and 4 of Algorithm 5.1), and this analysis is exact.

5.4.4 Numerical Comparison

In this section we test our approximate evaluation method. For instances with a small number of local warehouses ($|J| \leq 4$) we compare the approximate evaluation to exact evaluation. As we mentioned, evaluation is done for each SKU individually, so our instances each consider one SKU only. We assume that each local warehouse serves one group of machines, so we can represent demand rates as $M_{i,j}$, $j \in J$.

We first consider instances with main local warehouses only. Table 5.1 gives parameter settings and results for symmetric instances, and Tables 5.2 and 5.3 give parameter settings and results for asymmetric instances. Symmetric instances have equal demand rates and basestock levels in all mains, and have a cyclic pre-defined order for a lateral transshipment, so that each main observes the same demand process for lateral transshipments. Asymmetric instances have different demand rates, different basestock levels per location, or both, and for these asymmetric instances, we also vary the pre-defined lateral transshipments order. We study instances with 2 and 4 main local warehouses, and different values for demand rates $M_{i,k}$ and basestock levels $S_{i,k}$, $k \in K$. We have taken instances with basestock levels of at most 2. Under these low basestock levels, one gets low, medium and high fractions of demand streams that result in demands for lateral transshipments, and therefore this choice for the basestock levels is justified.

For instances with main local warehouses only, $\theta_{i,k}(\mathbf{S}_i)$, as determined in our approximate evaluation method, is equal for each $k \in K$. Furthermore, it is exact as it is obtained by an analysis of the aggregate system of all mains. Thus, the difference between the approximate and exact $\beta_{i,k}(\mathbf{S}_i)$ is equal to the difference between the exact and approximate $A_{i,k}(\mathbf{S}_i)$ (cf. (5.1)).

Table 5.1 Parameter settings and results for symmetric instances with mains only, with $M_{i,k}$ in parts per year, with $t = 0.04$ years, and with $\sigma(1) = (2)$ and $\sigma(2) = (1)$ for $|K| = 2$, and $\sigma(1) = (2,3,4)$, $\sigma(2) = (3,4,1)$, $\sigma(3) = (4,1,2)$ and $\sigma(4) = (1,2,3)$ for $|K| = 4$

$\|K\|$	$M_{i,k}$	$S_{i,k}$	$\beta_{i,k}(\mathbf{S}_i)$		$\alpha_{i,k,\sigma_1(k)}(\mathbf{S}_i)$		$\alpha_{i,k,\sigma_2(k)}(\mathbf{S}_i)$		$\alpha_{i,k,\sigma_3(k)}(\mathbf{S}_i)$		$\theta_{i,k}(\mathbf{S}_i)$
			Exact	Appr.	Exact	Appr.	Exact	Appr.	Exact	Appr.	
2	0.5	1	0.980	0.980	0.019	0.019					0.001
2	1	1	0.960	0.960	0.037	0.037					0.003
2	5	1	0.811	0.811	0.135	0.135					0.054
2	10	1	0.660	0.660	0.189	0.189					0.151
2	50	1	0.231	0.231	0.154	0.154					0.615
2	5	2	0.983	0.983	0.016	0.016					0.001
2	10	2	0.941	0.941	0.052	0.051					0.008
2	50	2	0.489	0.492	0.201	0.197					0.311
4	0.5	1	0.980	0.980	0.019	0.020	0.001	0.000	0.000	0.000	0.000
4	1	1	0.960	0.960	0.038	0.038	0.002	0.002	0.000	0.000	0.000
4	5	1	0.802	0.802	0.145	0.154	0.036	0.031	0.010	0.006	0.008
4	10	1	0.623	0.623	0.203	0.211	0.082	0.080	0.035	0.030	0.056
4	50	1	0.149	0.149	0.114	0.107	0.090	0.091	0.072	0.078	0.575
4	5	2	0.983	0.983	0.016	0.017	0.000	0.000	0.000	0.000	0.000
4	10	2	0.940	0.940	0.054	0.056	0.005	0.003	0.001	0.000	0.000
4	50	2	0.386	0.391	0.195	0.189	0.114	0.115	0.069	0.070	0.236

Table 5.2 Parameter settings for asymmetric instances with mains only, with $M_{i,k}$ in parts per year, with $t = 0.04$ years, and with $\sigma(1) = (2)$ and $\sigma(2) = (1)$ for $|K| = 2$, and for $|K| = 4$ either $\sigma(1) = (2,3,4)$, $\sigma(2) = (3,4,1)$, $\sigma(3) = (4,1,2)$ and $\sigma(4) = (1,2,3)$ (cycle) or $\sigma(1) = (2,3,4)$, $\sigma(2) = (1,3,4)$, $\sigma(3) = (1,2,4)$ and $\sigma(4) = (1,2,3)$ (dom., i.e., dominance)

Instance	$\|K\|$	$\sigma(k)$	$M_{i,1}$	$M_{i,2}$	$M_{i,3}$	$M_{i,4}$	$S_{i,1}$	$S_{i,2}$	$S_{i,3}$	$S_{i,4}$
1	2		1	5			1	1		
2	2		1	5			1	2		
3	2		5	10			1	1		
4	2		5	10			1	2		
5	4	Cycle	1	5	5	10	1	1	1	1
6	4	Cycle	1	5	5	10	1	1	1	2
7	4	Cycle	1	5	5	10	1	1	2	2
8	4	Cycle	1	5	5	10	1	2	2	2
9	4	Cycle	5	5	5	5	1	1	1	2
10	4	Cycle	5	5	5	5	1	2	2	2
11	4	Dom.	1	5	5	10	1	1	1	1
12	4	Dom.	1	5	5	10	1	1	1	2
13	4	Dom.	1	5	5	10	1	1	2	2
14	4	Dom.	1	5	5	10	1	2	2	2
15	4	Dom.	5	5	5	5	1	1	1	2
16	4	Dom.	5	5	5	5	1	2	2	2

For symmetric instances, we report $\beta_{i,k}(\mathbf{S}_i)$ as determined exactly and approximately, and $\theta_{i,k}(\mathbf{S}_i)$. Further, we study the lateral transshipment fraction $A_{i,k}(\mathbf{S}_i)$ on the disaggregated level by depicting $\alpha_{i,k,\tilde{k}}(\mathbf{S}_i)$ (specifying which main $\tilde{k} \in K$ provides the lateral transshipment to main k). Table 5.1 shows that the maximum absolute error for $\beta_{i,k}(\mathbf{S}_i)$ is 0.5 % (this value is obtained for the instance with

Table 5.3 Results for asymmetric instances with mains only

Instance	$\beta_{i,1}(\mathbf{S}_i)$		$\beta_{i,2}(\mathbf{S}_i)$		$\beta_{i,3}(\mathbf{S}_i)$		$\beta_{i,4}(\mathbf{S}_i)$		$\theta_{i,k}(\mathbf{S}_i)$
	Exact	Appr.	Exact	Appr.	Exact	Appr.	Exact	Appr.	
1	0.934	0.934	0.832	0.832					0.023
2	0.959	0.959	0.983	0.983					0.002
3	0.765	0.765	0.695	0.695					0.101
4	0.819	0.819	0.938	0.938					0.020
5	0.859	0.852	0.811	0.816	0.805	0.807	0.692	0.692	0.009
6	0.938	0.936	0.829	0.830	0.811	0.810	0.935	0.936	0.002
7	0.943	0.941	0.830	0.831	0.977	0.978	0.945	0.945	0.000
8	0.944	0.942	0.983	0.983	0.983	0.983	0.945	0.945	0.000
9	0.829	0.829	0.811	0.810	0.805	0.804	0.974	0.976	0.001
10	0.831	0.831	0.978	0.978	0.983	0.983	0.983	0.983	0.000
11	0.827	0.818	0.808	0.811	0.821	0.825	0.712	0.713	0.009
12	0.891	0.885	0.825	0.826	0.828	0.830	0.945	0.946	0.002
13	0.914	0.910	0.829	0.829	0.982	0.983	0.946	0.946	0.000
14	0.939	0.936	0.983	0.983	0.983	0.984	0.946	0.946	0.000
15	0.787	0.782	0.802	0.799	0.819	0.821	0.981	0.983	0.001
16	0.827	0.826	0.977	0.978	0.983	0.983	0.984	0.984	0.000

$|K| = 4$, $M_{i,k} = 50$, and $S_{i,k} = 2$) and the maximum absolute error for $\alpha_{i,k,\tilde{k}}(\mathbf{S}_i)$ is 0.9 % (see the instance with $|K| = 4$, $M_{i,k} = 5$, and $S_{i,k} = 1$).

For asymmetric instances, we report exact values for $\theta_{i,k}(\mathbf{S}_i)$ (equal for each $k \in K$), and exact and approximate values for $\beta_{i,k}(\mathbf{S}_i)$, $k \in K$. Errors in $A_{i,k}(\mathbf{S}_i)$, $k \in K$, follow immediately from Eq. (5.1) (for asymmetric instances, we do not study $\alpha_{i,k,\tilde{k}}(S_i)$). Table 5.3 shows us the results for instances as defined in Table 5.2. Again, we observe small deviations. Absolute errors are less than 1 % for all $\beta_{i,k}(\mathbf{S}_i)$, $k \in K$, in all instances.

So far, we considered instances with main local warehouses only. That means that we specifically tested our approximation step described in Sect. 5.4.2 that decouples the mains. From the numerical results, we can conclude that this approximation step works well. It results in small errors in performance measures compared to exact results.

In the remainder of this section, we describe instances having both main and regular local warehouses. For instances as described in Table 5.4, we report results in Table 5.5. We report results of the exact and approximate evaluation for mains and regulars separately. Notice that in our approximate evaluation, $\beta_{i,j}(\mathbf{S}_i)$ is determined exactly for the regulars $j \in J \setminus K$.

As can be seen from Table 5.5, again results from the approximate evaluation are close to exact results, although the maximum absolute error is now 2 %, obtained for $A_{i,j}(\mathbf{S}_i)$, and thus also for $\theta_{i,j}(\mathbf{S}_i)$, for the regulars in instance 6. This is higher than in instances with main local warehouses only. For the mains, the maximum absolute error is 1.1 %, obtained in instance 16. Notice that for instances with one main (instances 1–20), it is clear that the decoupling of the regulars from the main is the one and only cause of errors, since for one main, no decoupling of mains takes place. For the instances with 2 mains (instances 21–30), both decoupling of the regulars from the mains and decoupling of the system of mains do take place.

Table 5.4 Parameter settings for instances with mains and regulars, with $M_{i,k}$ in parts per year and with $t = 0.04$ years, with $K = \{1,2\}$, $\sigma(1) = (2)$ and $\sigma(2) = (1)$ for $|K| = 2$, with $k_j = 1$, $j \in J \setminus K$, for $|K| = 1$, and $J \setminus K = \{3,4\}$, $k_3 = 1$ and $k_4 = 2$ for $|K| = 2$

Instance	$\|K\|$	$\|J\| - \|K\|$	$M_{i,k}$ ($k \in K$)	$M_{i,j}$ ($j \in J \setminus K$)	$S_{i,k}$ ($k \in K$)	$S_{i,j}$ ($j \in J \setminus K$)
1	1	1	0.5	0.5	1	1
2	1	1	1	1	1	1
3	1	1	5	5	1	1
4	1	1	10	10	1	1
5	1	1	50	50	1	1
6	1	1	5	10	1	1
7	1	1	10	5	1	1
8	1	1	5	10	1	2
9	1	1	5	10	2	1
10	1	1	10	5	2	1
11	1	2	0.5	0.5	1	1
12	1	2	1	1	1	1
13	1	2	5	5	1	1
14	1	2	10	10	1	1
15	1	2	50	50	1	1
16	1	2	5	10	1	1
17	1	2	10	5	1	1
18	1	2	5	10	1	2
19	1	2	5	10	2	1
20	1	2	10	5	2	1
21	2	2	0.5	0.5	1	1
22	2	2	1	1	1	1
23	2	2	5	5	1	1
24	2	2	10	10	1	1
25	2	2	50	50	1	1
26	2	2	5	10	1	1
27	2	2	10	5	1	1
28	2	2	5	10	1	2
29	2	2	5	10	2	1
30	2	2	10	5	2	1

For our numerical comparison, we used instances with a few local warehouses only. Computation times were small; we measured computation times in milliseconds, and they were mostly zero. However, for the exact evaluation, some instances required 10 ms. So for those instances there is a large difference in computation time between the exact and the approximate evaluation. This difference will even be larger for instances with a larger number of local warehouses, and as $|J|$ becomes too large, exact evaluation will be impossible, due to the size of the state spaces.

In summary, we can conclude that our approximate evaluation is accurate and fast.

Table 5.5 Results for instances with mains and regulars

Inst	$\beta_{i,k}(\mathbf{S}_i), k \in K$		$\beta_{i,j}(\mathbf{S}_i)$	$A_{i,k}(\mathbf{S}_i), k \in K$		$A_{i,j}(\mathbf{S}_i), j \in J \setminus K$		$\theta_{i,k}(\mathbf{S}_i), k \in K$		$\theta_{i,j}(\mathbf{S}_i), j \in J \setminus K$	
	Exact	Appr.	$(j \in J \setminus K)$	Exact	Appr.	Exact	Appr.	Exact	Appr.	Exact	Appr.
1	0.980	0.980	0.980			0.019	0.019	0.0200	0.0200	0.0006	0.0004
2	0.960	0.960	0.962			0.036	0.037	0.0399	0.0399	0.0022	0.0015
3	0.812	0.811	0.833			0.127	0.135	0.1878	0.1892	0.0401	0.0315
4	0.665	0.660	0.714			0.172	0.189	0.3350	0.3396	0.1133	0.0970
5	0.238	0.231	0.333			0.143	0.154	0.7619	0.7692	0.5238	0.5128
6	0.767	0.761	0.714			0.198	0.217	0.2325	0.2391	0.0881	0.0683
7	0.699	0.698	0.833			0.109	0.116	0.3013	0.3023	0.0572	0.0504
8	0.820	0.819	0.946			0.040	0.044	0.1798	0.1814	0.0145	0.0098
9	0.962	0.964	0.714			0.267	0.275	0.0382	0.0362	0.0185	0.0103
10	0.938	0.939	0.833			0.153	0.156	0.0617	0.0615	0.0135	0.0102
11	0.980	0.980	0.980			0.019	0.019	0.0204	0.0204	0.0006	0.0004
12	0.959	0.959	0.962			0.036	0.037	0.0413	0.0413	0.0023	0.0016
13	0.792	0.789	0.833			0.124	0.132	0.2079	0.2105	0.0431	0.0351
14	0.622	0.614	0.714			0.162	0.175	0.3781	0.3860	0.1240	0.1103
15	0.184	0.176	0.333			0.112	0.118	0.8162	0.8235	0.5545	0.5490
16	0.711	0.700	0.714			0.184	0.200	0.2891	0.3000	0.1020	0.0857
17	0.684	0.682	0.833			0.107	0.114	0.3163	0.3182	0.0595	0.0530
18	0.807	0.804	0.946			0.039	0.043	0.1926	0.1957	0.0151	0.0106
19	0.937	0.940	0.714			0.259	0.268	0.0627	0.0604	0.0265	0.0173
20	0.931	0.931	0.833			0.152	0.155	0.0694	0.0691	0.0149	0.0115
21	0.980	0.980	0.980	0.020	0.020	0.020	0.020	0.0008	0.0008	0.0000	0.0000
22	0.959	0.959	0.962	0.038	0.038	0.038	0.038	0.0032	0.0032	0.0002	0.0001
23	0.784	0.783	0.833	0.147	0.148	0.152	0.155	0.0694	0.0691	0.0149	0.0115
24	0.595	0.592	0.714	0.199	0.201	0.217	0.227	0.2055	0.2068	0.0688	0.0591
25	0.149	0.145	0.333	0.113	0.112	0.164	0.171	0.7382	0.7435	0.5028	0.4957
26	0.724	0.720	0.714	0.168	0.172	0.245	0.255	0.1085	0.1082	0.0408	0.0309
27	0.640	0.639	0.833	0.193	0.193	0.135	0.139	0.1673	0.1675	0.0316	0.0279
28	0.793	0.793	0.946	0.143	0.144	0.049	0.051	0.0641	0.0637	0.0054	0.0034
29	0.958	0.962	0.714	0.037	0.035	0.284	0.285	0.0044	0.0035	0.0021	0.0010
30	0.931	0.932	0.833	0.059	0.058	0.164	0.165	0.0101	0.0099	0.0022	0.0017

5.5 Greedy Heuristic

In the previous section, we have described how, for each SKU $i \in I$, a given policy, i.e., a choice for all basestock levels $S_{i,j}$, $j \in J$, can be evaluated in an approximate way. The current section deals with a heuristic algorithm to find a feasible policy for Problem (P) with as low costs as possible. For this integer-programming problem with non-linear constraints, we provide a greedy heuristic, in which we use our approximate evaluation method to determine the average costs and mean waiting times of given policies.

Wong et al. [30] have shown that with exact evaluation, the greedy algorithm followed by local search performs very well (they study instances with main local warehouses only and one customer group per local warehouse). We carried out additional experiments with the greedy method and exact evaluation, but without local search. For instances with 50 and 100 SKU's, and 2, 3, or 4 local warehouses, all of which are mains, we found gaps of at most 3.7 % compared to a lower bound obtained by Dantzig-Wolfe decomposition, and the size of this gap was decreasing in the number of SKU's: the average gap for instances with 50 SKU's was 1.00 % and the average gap for instances with 100 SKU's was 0.72 %. This shows that the greedy heuristic performs well.

Our greedy heuristic is similar to the greedy heuristic for the problem of Chap. 3; see Algorithm 3.1. This heuristic consists of three steps. In the first (initialization) step, we set all basestock levels $S_{i,j} := 0$, $i \in I$, $j \in J$. In the second step, we increase basestock levels if and as long as it decreases (i.e., does not increase) costs. Obviously, increasing basestock levels has at the same time a decreasing effect on the waiting times. We execute this step for each SKU $i \in I$ separately (since costs for SKU i depend on $\mathbf{S}_i$ only). Notice that the change in costs $C_i(\mathbf{S}_i)$ for an SKU i when $S_{i,j}$ is increased by 1 unit for some $j \in J$ is given by

$$\Delta_j C_i(\mathbf{S}_i) = C_i(\mathbf{S}_i + \mathbf{e}_j) - C_i(\mathbf{S}_i), \tag{5.10}$$

where $\mathbf{e}_j$ is the j-th unit vector with dimension $|I|$. As long as $\Delta_j C_i(\mathbf{S}_i) \leq 0$ for some $i \in I$ and $j \in J$, we increase the basestock level that gives us the largest costs decrease. These first two steps are similar to Step 1 of Algorithm 3.1.

In the third step, as long as our current solution is not feasible, we iteratively increase the basestock levels $S_{i,j}$, $i \in I$, $j \in J$. Here we follow the same logic as in Step 2 of Algorithm 3.1. We divide the set $\mathscr{S}$ of all solutions into a subset

$$\mathscr{S}^{\text{feas}} := \{\mathbf{S} \in \mathscr{S} | \hat{W}_n(\mathbf{S}) \leq \hat{W}_n^{\text{obj}} \text{ for all } j \in J, n \in N_j\}$$

of feasible solutions and the subset $\mathscr{S} \setminus \mathscr{S}^{\text{feas}}$ of non-feasible solutions. Next, for each solution $\mathbf{S} \in \mathscr{S}$, we define the distance $d(\mathbf{S})$ to the set $\mathscr{S}^{\text{feas}}$ of feasible solutions as

$$\begin{aligned} d(\mathbf{S}) &:= \sum_{n \in N} \left(\hat{W}_n(\mathbf{S}) - \hat{W}_n^{\text{obj}}\right)^+ \\ &= \sum_{j \in J} \sum_{n \in N_j} \left(\sum_{i \in I} \frac{m_{i,n}}{M_n} W_{i,j}(\mathbf{S}_i) - \hat{W}_n^{\text{obj}}\right)^+, \end{aligned}$$

with $x^+ := \max\{0, x\}$ for all $x \in \mathbb{R}$. Next, in each greedy step, we have a current solution $\mathbf{S}$, and we look at the ratio of the decrease in distance to the set of feasible solutions and the increase in costs if one of the basestock levels $S_{i,j}$ is increased by one unit. Let $i \in I$ and $j \in J$, and suppose that $S_{i,j}$ would be increased by 1 unit. Then the increase in costs is equal to

$$\Delta_{i,j} C(\mathbf{S}) = \Delta_j C_i(\mathbf{S}_i) = C_i(\mathbf{S}_i + \mathbf{e}_j) - C_i(\mathbf{S}_i), \tag{5.11}$$

(because we first execute the steps 1 and 2, all $\Delta_{i,j} C(\mathbf{S}) > 0$ in each greedy step). The decrease in distance to the feasible solutions is given by $-\Delta_{i,j} d(\mathbf{S})$, where ($\mathbf{E}_{i,j}$ is an $|I| \times |J|$ matrix with a 1 on the (i,j)-th position and zeros on all other positions)

$$
\begin{aligned}
\Delta_{i,j}d(\mathbf{S}) &= d(\mathbf{S}+\mathbf{E}_{i,j}) - d(\mathbf{S}) \\
&= \sum_{j'\in J}\sum_{n\in N_{j'}} \left[\left(\hat{W}_n(\mathbf{S}+\mathbf{E}_{i,j}) - \hat{W}_n^{\mathrm{obj}}\right)^+ - \left(\hat{W}_n(\mathbf{S}) - \hat{W}_n^{\mathrm{obj}}\right)^+ \right]. \\
&= \sum_{j'\in J}\sum_{n\in N_{j'}} \left[\left(\sum_{i'\in I\setminus\{i\}} \frac{m_{i',n}}{M_n} W_{i',j'}(\mathbf{S}_{i'}) + \frac{m_{i,n}}{M_n} W_{i,j'}(\mathbf{S}_i+\mathbf{e}_j) - \hat{W}_n^{\mathrm{obj}} \right)^+ \right. \\
&\qquad \left. - \left(\sum_{i'\in I} \frac{m_{i',n}}{M_n} W_{i',j'}(\mathbf{S}_{i'}) - \hat{W}_n^{\mathrm{obj}} \right)^+ \right].
\end{aligned}
\tag{5.12}
$$

Notice that increasing $S_{i,j}$ has a decreasing effect on the mean waiting times $W_{i,j}(\mathbf{S}_i)$ (this may be shown by a sample path reasoning), and no effect on the mean waiting times for other SKU's. By (5.12), it is then easily verified that $\Delta_{i,j}d(\mathbf{S}) \geq 0$. The factor

$$
\Gamma_{i,j} = \frac{-\Delta_{i,j}d(\mathbf{S})}{\Delta_{i,j}C(\mathbf{S})} \tag{5.13}
$$

denotes the decrease in distance to the feasible solutions per unit of increase in costs. For the combination of SKU i and local warehouse j with the highest ratio $\Gamma_{i,j}$, we have the "biggest bang for the buck" and we increase the basestock level to $S_{i,j}+1$. We continue with the above greedy steps until we arrive at a feasible solution. The formal procedure is described in Algorithm 5.3.

Algorithm 5.3 *(Greedy heuristic)*

Step 1 Set $S_{i,j} := 0,\ i \in I,\ j \in J$
Step 2 For each SKU $i \in I$:

- Step 2-a Calculate $\Delta_j C_i(\mathbf{S}_i),\ j \in J$ (cf. (5.10));
- Step 2-b While $\min_{j\in J}\{\Delta_j C_i(\mathbf{S}_i)\} \leq 0$:
  - (a) Determine j' such that $\Delta_{j'} C_i(\mathbf{S}_i) \leq \Delta_j C_i(\mathbf{S}_i),\ j \in J$;
  - (b) Set $S_{i,j'} := S_{i,j'} + 1$;
  - (c) Calculate $\Delta_j C_i(\mathbf{S}_i),\ j \in J$

Step 3

- Step 3-a Calculate $\Delta_{i,j}C(\mathbf{S})$, $\Delta_{i,j}d(\mathbf{S})$, and $\Gamma_{i,j},\ i \in I,\ j \in J$ (cf. (5.11)–(5.13))
- Step 3-b While $d(\mathbf{S}) > 0$:
  - (a) Determine i' and j' such that $\Gamma_{i',j'} \geq \Gamma_{i,j},\ i \in I,\ j \in J$;
  - (b) Set $S_{i',j'} := S_{i',j'} + 1$;
  - (c) Calculate $\Delta_{i,j}C(\mathbf{S})$, $\Delta_{i,j}d(\mathbf{S})$, and $\Gamma_{i,j},\ i \in I,\ j \in J$

Computation time can be saved in the execution of the algorithm if only those (intermediate) results are updated that are affected by a certain change. E.g., waiting times $W_{i,j}(\mathbf{S}_i)$, $j \in J$, only depend on the basestock levels for SKU i, and are not subject to change if a basestock level for another SKU is increased.

5.6 Partial vs. Full Pooling

In this section we investigate numerically how partial pooling performs compared to the full pooling situation in terms of costs. We use the greedy heuristic described in the previous section, where evaluation of given policies is done approximately, as described in Sect. 5.4.

We use a small data set with 5 local warehouses. Each local warehouse serves one group of machines. In our data set we have 50 SKU's with fictitious cost data and equal demand rates per group. Cost prices for SKU's 1, 2, …, 50, are 2,000, 4,000, …, 100,000 Euros, respectively. The inventory holding cost parameter for holding one item on stock for one day is calculated as $0.25/365$ times the SKU price. Demand rates per group per day for SKU's 1, 2, …, 50 are 0.0100, 0.0098, …, 0.0002, respectively. Since each warehouse serves one group of machines, the demand rate per day for an SKU i represents both $m_{i,n}$, $n \in N$, and $M_{i,j}$, $j \in J$.

Target aggregate mean waiting times are equal for all groups at the different local warehouses, and set equal to 0.10 days. All $t^{\text{lat}}_{j,k}$, $j \in J$, $k \in K$, are equal to 0.5 days. Further, all t^{em}_{j}, $j \in J$, are equal to 2 days and $t = 14$ days. All $c^{\text{lat}}_{j,k}$, $j \in J$, $k \in K$ are equal to 500 Euros, and all c^{em}_{j}, $j \in J$, are equal to 1,000 Euros.

For the data set described above, we study 6 different cases, where the number of main local warehouses $|K|$ varies from 0 to 5. In each case, the other $5 - |K|$ local warehouses are regulars. We use a cyclic lateral transshipment order, and assignments of regulars to mains are distributed over the mains in an as equal as possible way. E.g., for the case with $|K|$=2 mains $\{1,2\}$ and 3 regulars $\{3,4,5\}$, we then have $\sigma(1) = (2)$, $\sigma(2) = (1)$, $k_3 = k_5 = 1$, and $k_4 = 2$.

For these six cases, we report the total yearly costs for the solution obtained by the greedy algorithm in Table 5.6. Furthermore, we report the cost savings compared to the no pooling case with $|K| = 0$. Compared to the no pooling situation, the full pooling situation implies a cost savings of 35.1 %. A very important observation here is that if we apply partial pooling with *one* main local warehouse(s) only, already 21.9/35.1 = 62.3 % of the full pooling benefits are obtained. This is an important observation. With two main local warehouses, we obtain already 31.1/35.1 = 88.7 % of the full pooling benefits. As we mentioned in Sect. 5.1, because of practical considerations it is often not desirable to let all local warehouses carry out lateral transshipments. Also, the operational costs may be too high to make all local warehouses main local warehouses. Our results imply that a major part of the possible benefits of using lateral transshipments is already obtained by a low number of mains.

In addition to the above results, we can say the following on the use of approximate evaluations in the greedy heuristic:

(a) We evaluated the final solution obtained by the greedy algorithm in each of the six cases in Table 5.6 exactly. The differences in mean waiting time are small compared to the waiting times that we obtained by approximate evaluation: The waiting time values that we obtain deviate at most 1.52 % from the exact values.

Table 5.6 Results for data set with 50 SKU's for cases with different $|K|$

$\|K\|$	Total yearly cost (Euros)	Cost savings compared to no pooling (%)
0	$2.80 \cdot 10^6$	
1	$2.19 \cdot 10^6$	21.9
2	$1.93 \cdot 10^6$	31.1
3	$1.89 \cdot 10^6$	32.7
4	$1.82 \cdot 10^6$	35.1
5	$1.82 \cdot 10^6$	35.1

(b) We ran the greedy heuristic using exact evaluation instead of approximate evaluation. Although the final solution in some of the six cases differs slightly, the total costs are almost equal to the outcomes in Table 5.6: The values for total costs as listed in Table 5.6 deviate at most 2.02 % from the total costs obtained by the greedy heuristic with exact evaluations.
(c) The greedy algorithm with exact evaluation on average needed 163 s, where the results in Table 5.6, using our greedy algorithm with approximate evaluation, are obtained much quicker: in 30 ms on average.

In Sect. 5.4, we have shown that our approximate evaluation method obtains accurate results, i.e., results that are reasonably close to values as obtained by exact evaluation. The results obtained in this section confirm that the approximate evaluation method is accurate. Furthermore, the gain in computation time is evident.

5.7 Case Study: ASML

In this section, we report on a case study carried out at ASML in 2004. We received data for all 19 local warehouses in the US. In Sects. 5.4 and 5.6, we have shown numerically that our approximate evaluation method as described in Sect. 5.4 is accurate. We will use this evaluation method here. The aim of this section is to show potential savings of pooling spare parts inventory at ASML. In line with the main result in Sect. 5.6, we will see that with partial pooling already a substantial part of the full pooling benefits can be obtained.

In this section, we first describe the ASML data set. Then, we compare the total costs for cases with different numbers of mains. Finally, we spend attention to savings compared to the old situation at ASML, i.e., the situation that lateral transshipment was not taken into account in the planning, but nevertheless used in practice.

The data set for the 19 local warehouses in the United States of America is as follows. We have 1,451 SKU's, where the price of the most expensive SKU is about 10^6 times the price of the cheapest SKU. Inventory holding cost rates per year are 0.25 times the price. All $t^{\text{lat}}_{j,k}$, $j \in J$, $k \in K$, are equal to 0.5 days. Further, $t^{\text{em}}_j = 2$ days for all $j \in J$ and $t = 14$ days. All $c^{\text{lat}}_{j,k}$, $j \in J$, $k \in K$ are equal to 500 Euros, and $c^{\text{em}}_j = 1{,}000$ Euros for all $j \in J$.

Table 5.7 Results for case study at ASML for cases with different $|K|$

$\|K\|$	Total yearly cost (normalized)	Cost savings compared to no pooling (%)	Computation time (s)
0	100		71
4	51.82	48.18	57
19	49.81	50.19	132

In our case, in total 27 groups of machines are present, and each of the 19 local warehouses serves one or two groups. Different groups at a local warehouse correspond to different machine types, which have several common components. For each of the groups $n \in N$, the target aggregate mean waiting time $\hat{W}_n^{\text{obj}}$ is set at 0.15 days. Demand rates are low; the highest value for $m_{i,n}$ that occurs is 39 units per year.

In the ASML data set, four local warehouses are identified by ASML as suitable candidates to be mains. A pre-specified order for lateral transshipment between these mains is given by ASML, as well as k_j for all regular local warehouses $j \in J \setminus K$.

To compare the total yearly costs at different numbers of main local warehouses, we study cases with $|K| = 0, 4$, and 19, respectively. At $|K| = 19$, as pre-specified lateral transshipment order we assume that first the four local warehouses are checked that are mains in the case with $|K| = 4$, and then all other local warehouses. Results are given in Table 5.7. The total yearly costs in the no pooling situation is normalized to 100. Table 5.7 shows that in the full pooling case more than 50 % of the no pooling costs can be saved. In line with results obtained in the previous section, we observe that in the partial pooling situation with four mains, the majority of these potential savings can be obtained: 48.18/50.19 = 96.0 %. Clearly, the additional value of having more main local warehouses is very small. Computation time results show that the algorithm is fast enough.

While the results in Table 5.7 show that pooling is important, we needed a second experiment to verify the potential savings of the new model for ASML. Until 2004, ASML did not take lateral transshipment into account at the tactical planning level, i.e., the inventory in each local warehouse was planned separately, but they did use lateral transshipments at the operational level. Given the parameter values for transportation times and costs at ASML, this meant that in practice, ASML obtained lower waiting times than it planned for, since the actual performance was better because of the lateral transshipments used at the operational level. To make a fair comparison between the old planning model and the planning model of this chapter, we made the following comparison (in both models, we work with four main local warehouses):

- *Old planning model*: At the tactical level, we determine the basestock levels per local warehouse (by using the model of Chap. 3, or, equivalently, the model of the current chapter with $|J| = |K| = 1$). The target aggregate mean waiting times are set at 0.15 days. Next, the performance under the given basestock levels for the whole system is evaluated by the approximate evaluation method of this chapter.

Table 5.8 Results for case study at ASML: comparison between old and new model at $|K| = 4$

Model	Target waiting time in days (avg)	Realized waiting time in days (avg)	Total yearly cost (normalized)
Old model	0.1500	0.1007	100
New model	0.1007	0.0975	68.51

This results in the actual performance, in terms of aggregate mean waiting times and total costs. The total costs were normed at 100. The realized aggregate mean waiting times were used as targets for the new model.

- *New planning model*: This is the model as described in this chapter, which is run with the waiting time targets stemming from the actual performance under the old model. The costs under the solution generated by the new model appear to be 31.5 % lower. This value denotes the efficiency improvement that ASML could expect by incorporating lateral transshipments at the tactical planning level.

The results of this experiment are also listed in Table 5.8.

The large difference in results under the old and new model, respectively, is explained by a more balanced choice of basestock levels under the new model. Due to the low demand rates at ASML, almost all basestock levels at the local warehouse are equal to 0 or 1. By a planning per local warehouse, the SKU's are, in a way, ranked on the basis of the ratio of cost price over demand rate, and only the SKU's with the lowest ratios are placed on stock. Because rankings at different local warehouses can be expected to be similar, this leads to many overlaps between the packages of SKU's that are placed on stock in the various local warehouses. Under the new model, one obtains a more balanced solution. SKU's with the lowest ratios may be placed on stock in all local warehouses, and SKU's with very high ratios may not be placed on stock at all, as under the old model. But for many SKU's in between, the number of warehouses where a SKU is placed on stock is a balanced number between 1 and $|J| - 1 = 18$. The lower the ratio the higher the number of warehouses where the SKU is placed in stock. In this way, one gets complementary stocks, and that is what one needs when using lateral transshipments.

Based on the above case study, ASML decided to implement the greedy heuristic for the tactical planning of the spare parts inventory in all local warehouses in their world-wide service network. ASML is using the algorithm since early 2006 and succeeded in reducing *both waiting times and costs* considerably. Obviously, the implementation of this new planning method is not a stand-alone project, and the results have to be seen in the light of a combination of projects. The performance improvement has been large. ASML has been measuring the performance of an older generation of machines. In the years 2006–2010, the service rates have improved by a factor 10, and the total inventory decreased by a factor two! The savings, using the greedy heuristic, have been used to reallocate the inventories and this has led to an improved customer service. In the meantime, services and spare part provision have become value drivers and distinguishing competitive factors (see [9] for further details about this implementation).

5.8 Concluding Remarks

As stated in the introduction of this chapter, the model and results of this chapter stem from [12]. The model is a multi-item, multi-location inventory model with lateral transshipments and aggregate mean waiting time constraints. It is because of these system-oriented service constraints that one obtains a multi-item model. The first studies on such models were executed by Wong et al. [30, 31]. An earlier multi-item model with two locations, lateral transshipments and a joint storage constraint was studied by Archibald et al. [3].

The multi-item model is such that the evaluation of a given basestock policy decomposes into an evaluation of single-item instances, and thus one can build on methods for single-item problems with lateral transshipments. In the approximate evaluation method as presented in this chapter, we first reduce the problem to a system with main local warehouses, and the latter system is evaluated by defining Poisson overflow streams and steady-state analysis of single stockpoints. This powerful basic idea stems from Axsäter [4], who used it for a two-echelon system with lateral transshipments but without emergency shipments. It has been used and tailored for several models with different assumptions regarding the lateral transshipments and emergency shipments; see Alfredsson and Verrijdt [1], Kukreja et al. [13], Kutanoglu [14], and Reijnen et al. [19]. Generally, an accurate approximate evaluation method is obtained. Nevertheless, for some systems, it may be relevant to model the overflow streams more accurately. A more accurate method can be obtained by modeling overflow streams as on-off Poisson streams; see Van Wijk et al. [26].

The model of this chapter has a partial pooling structure, with main and regular local warehouses. This structure has full pooling and no pooling as special cases. A special case is obtained when one has one main local warehouse, which may model systems in practice with a so-called quick response warehouse. Such quick response warehouses are used at Volvo Parts Corporation (see Axsäter et al. [7] and Howard et al. [11]) and Océ Technologies (see Rijk [20] and Van Wijk et al. [28]). In the literature several other forms of partial pooling have been studied. For systems with pooling groups, where one has full pooling within groups of local warehouses and no pooling between different groups, see Lee [15] and Axsäter [4]. Another form of pooling is obtained when all local warehouses are ranked and lateral transshipments are only allowed from a warehouse to other warehouses with a lower rank. This leads to unidirectional lateral transshipments; see Axsäter [5] and Olsson [17] (this form has a close connection to inventory problems with demand substitution, see e.g. [8]). A general partial pooling structure, that generalizes all above forms, has been studied by Reijnen et al. [19].

Under the above structures, the pooling is partial because lateral transshipments are only allowed from certain local warehouses to certain other warehouses. One can also have rules under which local warehouses are only allowed to provide a lateral transshipments if their inventory level is above a certain threshold, called the hold back level. Such a rule avoids that the last part(s) in a local warehouse is (are) taken away while there is a high chance that you need them for a demand at the same warehouse in the time until a next replenishment order arrives; see e.g. Tagaras and

Cohen [23], Axsäter [6], Evers [10], Minner et al. [16], and Xu et al. [32]. For a combination of the general partial pooling structure of Reijnen et al. [19] and hold back levels, see Van Wijk et al. [27]. The use of hold back levels also relates to dynamic and optimal lateral transshipment policies; see e.g. Zhao et al. [33, 34], Van Wijk [26] (Chap. 4), and Tiemessen et al. [24]. One may also limit the use of lateral transshipments to certain SKU's and locations, and, in the context of two demand classes, to the high-priority demands; see Alvarez et al. [2]. Instead of, or in addition to, using reactive lateral transshipments, one can also consider proactive lateral transshipments; see Seidscher and Minner [22] for a comparison between reactive and proactive lateral transshipments. For an extensive, recent overview on inventory systems with reactive and proactive lateral transshipments, see Paterson et al. [18].

When it comes to implementations of lateral transshipments models in practice, besides the implementation at ASML, we can also report an implementation at Océ Technologies. Rijk [20] applied the model of this chapter at Océ to the spare parts stocks in the cars of service engineers and the quick response warehouses. The model is applied per region where one has one quick response warehouse, which represents a main local warehouse, and multiple car stocks (regular local warehouses). Rijk identified a huge potential improvement in terms of the first-day fix rate (percentage of problems that is solved on the same day) and total costs. After a successful pilot in France and a further refinement of the planning model, it has been rolled out in multiple countries in Europe.

Problems

5.1. Consider the instance in Table 5.1 with $|K| = 2$, $M_{i,k} = 5$, and $S_{i,k} = 1$.

(a) Depict for this instance the Markov process as described in Sect. 5.3. Give the Markov process that is obtained *before* uniformization is applied. (The uniformization step is applied to facilitate the numerical solution via a computer program. Here, the steady-state equations can be solved by hand.)
(b) Determine the steady-state probabilities $\pi(\mathbf{x}_i)$.
(c) Determine $\beta_{i,1}(\mathbf{S}_i)$, $\beta_{i,2}(\mathbf{S}_i)$, $\alpha_{i,1,2}(\mathbf{S}_i)$, $\alpha_{i,2,1}(\mathbf{S}_i)$, $\theta_{i,1}(\mathbf{S}_i)$, and $\theta_{i,2}(\mathbf{S}_i)$. Check whether your outcomes match with the results listed in Table 5.1.

5.2. Consider instance 6 in Table 5.4.

(a) Depict for this instance the Markov process as described in Sect. 5.3. Give the Markov process that is obtained *before* uniformization is applied.
(b) Determine the steady-state probabilities $\pi(\mathbf{x}_i)$.
(c) Determine $\beta_{i,1}(\mathbf{S}_i)$, $\beta_{i,2}(\mathbf{S}_i)$, $\alpha_{i,2,1}(\mathbf{S}_i)$, $\theta_{i,1}(\mathbf{S}_i)$, and $\theta_{i,2}(\mathbf{S}_i)$. Check whether your outcomes match with the results listed in Table 5.5.
(d) Determine $\beta_{i,1}(\mathbf{S}_i)$, $\beta_{i,2}(\mathbf{S}_i)$, $\alpha_{i,2,1}(\mathbf{S}_i)$, $\theta_{i,1}(\mathbf{S}_i)$, and $\theta_{i,2}(\mathbf{S}_i)$ via the approximate evaluation method as given in Sect. 5.4.3. Check whether your outcomes match with the results listed in Table 5.5.

5.3. Suppose that we have a situation in practice where the mean replenishment leadtime t, the mean emergency shipment times t_j^{em} and the mean lateral transshipment times $t_{j,k}^{\mathrm{lat}}$ are SKU-dependent, e.g., because different transport modes are used for different subsets of SKU's. For that situation, we can formulate an extended model with replenishment leadtimes t_i, mean emergency shipment times $t_{i,j}^{\mathrm{em}}$, and mean lateral transshipment times $t_{i,j,k}^{\mathrm{lat}}$.

(a) Denote which formulas in Sect. 5.2 change for the extended model and give the new formulas.
(b) Explain why both the exact and approximate evaluation of Sects. 5.3 and 5.4 will still work.
(c) What needs to be changed in the greedy heuristic for the extended model?

5.4. In the model of this chapter, it is assumed that inventory holding costs are paid for both parts on hand per local warehouse and parts in its replenishment pipeline; see Sect. 5.2. Suppose that we have a situation in practice where inventory holding costs are only paid for the parts on hand.

(a) Determine the mean number of parts in the replenishment pipeline of a local warehouse $j \in J$.
(b) Formula (5.4) for $C_i(\mathbf{S}_i)$ has to be adapted when inventory holding costs are only paid for the parts on hand. Give the adapted formula.
(c) Explain why Problem (P) remains the same in essence.

5.5. Consider the special case of the model described in this chapter with only main local warehouses. Suppose, $\sigma(1) = (2,3)$, $\sigma(2) = (3,1)$, and $\sigma(3) = (1,2)$. This implies that we have a symmetric structure for the lateral transshipments. For the other parameters, we also assume symmetric parameters, and we limit ourselves to a case with only one SKU. The demand rate per local warehouse is denoted by m (>0). The basestock level for each local warehouse is given by $S = 1$. Further, the mean replenishment leadtime t is equal to 1 time unit. We are interested in:

- β: The fill rate at a main local warehouse;
- θ: The fraction of demand at a main local warehouse that is satisfied via an emergency shipment;
- A: The fraction of demand at a main local warehouses that is satisfied via a lateral transshipment.

For this specific system, we compare exact and approximate evaluation as described in Sects. 5.3 and 5.4.

(a) The approximate evaluation procedure is described by Algorithms 5.1 and 5.2 in Sect. 5.4. The procedure generates a solution for Eqs. (5.5) and (5.6) and the other equations that are used in Algorithms 5.1 and 5.2. In the special case that is considered in this problem, this set of equations simplifies considerably. Formulate the simplified set of equations. In your notation, use $\hat{M}_1$ and $\hat{M}_2$ for the rate with which demands for a lateral transshipment are placed by a main local warehouse at the first and second alternative warehouse, respectively, and use $\hat{M}$ for the total demand rate at a main local warehouse (consisting of the external demand rate and overflow demand rates).

(b) Derive analytical expressions for β, θ, and A as obtained via the approximate evaluation procedure.

(c) Derive exact expressions for β, θ, and A.

(d) Determine the differences in the outcomes of (b) and (c).

References

1. Alfredsson, P., Verrijdt, J.: Modeling emergency supply flexibility in a two-echelon inventory system. Manag. Sci. **45**, 1416–1431 (1999)
2. Alvarez, E.M., Van der Heijden, M.C., Vliegen, I.M.H., Zijm, W.H.M.: Service differentiation through selective lateral transshipments. Eur. J. Oper. Res. **237**, 824–835 (2014)
3. Archibald, T.W., Sassen, S.A., Thomas, L.C.: An optimal policy for a two depot inventory problem with stock transfer. Manag. Sci. **43**, 173–183 (1997)
4. Axsäter, S.: Modeling emergency lateral transshipments in inventory systems. Manag. Sci. **36**, 1329–1338 (1990)
5. Axsäter, S.: Evaluation of unidirectional lateral transshipments and substitutions in inventory systems. Eur. J. Oper. Res. **149**, 438–447 (2003)
6. Axsäter, S.: A new decision rule for lateral transshipments in inventory systems. Manag. Sci. **49**, 1168–1179 (2003)
7. Axsäter, S., Howard, C., Marklund, J.: A distribution inventory model with transshipments from a support warehouse. IIE Trans. **45**, 309–322 (2013)
8. Deflem, Y., Van Nieuwenhuyse, I.: Managing inventories with one-way substitution: a Newsvendor analysis. Eur. J. Oper. Res. **228**, 484–493 (2013)
9. European Supply Chain Forum (Stein, W.): Spare parts planning at ASML. ESCF Operations Practices. http://www.escf.nl/operation (2012)
10. Evers, P.T.: Heuristics for assessing emergency transshipments. Eur. J. Oper. Res. **129**, 311–316 (2001)
11. Howard, C., Marklund, J., Tan, T., Reijnen, I.C.: Inventory control in a spare parts distribution system with emergency stocks and pipeline information. Manuf. Serv. Oper. Manag. http://dx.doi.org/10.1287/msom.2014.0508 (2015, to appear)
12. Kranenburg, A.A., Van Houtum, G.J.: A new partial pooling structure for spare parts networks. Eur. J. Oper. Res. **199**, 908–921 (2009)[1]
13. Kukreja, A., Schmidt, C.P., Miller, D.M.: Stocking decisions for low-usage items in a multi-location inventory system. Manag. Sci. **47**, 1371–1383 (2001)
14. Kutanoglu, E.: Insights into inventory sharing in service parts logistics systems with time-based service levels. Comput. Ind. Eng. **54**, 341–358 (2008)
15. Lee, H.L.: A multi-echelon inventory model for repairable items with emergency lateral transshipments. Manag. Sci. **33**, 1302–1316 (1987)
16. Minner, S., Silver, E.A., Robb, D.J.: An improved heuristic for deciding on emergency transshipments. Eur. J. Oper. Res. **148**, 384–400 (2003)
17. Olsson, F.: An inventory model with unidirectional lateral transshipments. Eur. J. Oper. Res. **200**, 725–732 (2010)
18. Paterson, C., Kiesmüller, G., Teunter, R., Glazebrook, K.: Inventory models with lateral transshipments: a review. Eur. J. Oper. Res. **210**, 125–136 (2011)

[1] This chapter has been reprinted from Kranenburg and Van Houtum [12] with kind permission from Elsevier.

19. Reijnen, I.C., Tan, T., Van Houtum, G.J.: Inventory planning for spare parts networks with delivery time requirements. Working paper 280, Beta Research School, Eindhoven University of Technology (2009)
20. Rijk, P.J.: Multi-item, multi-location stock control with capacity constraints for the fieldstock of service parts at Océ. MSc thesis, master program Industrial Engineering, Eindhoven University of Technology. http://w3.tue.nl/en/services/library/digilib/publications_from_tue/masters_theses/ (2007)
21. Ross, S.M.: Stochastic Processes. Wiley, New York (1996)
22. Seidscher, A., Minner, S.: A semi-Markov decision problem for proactive and reactive transshipments between multiple warehouses. Eur. J. Oper. Res. **230**, 42–52 (2013)
23. Tagaras, G., Cohen, M.A.: Pooling in two-location inventory systems with non-negligible replenishment lead times. Manag. Sci. **38**, 1067–1083 (1992)
24. Tiemessen, H.G.H., Fleischmann, M., Van Houtum, G.J., Van Nunen, J.A.E.E., Pratsini, E.: Dynamic demand fulfillment in spare parts networks with multiple customer classes. Eur. J. Oper. Res. **228**, 367–380 (2013)
25. Tijms, H.C.: A First Course in Stochastic Models. Wiley, Chichester (2003)
26. Van Wijk, A.C.C.: Pooling and polling: creation of pooling in inventory and queueing models. Ph.D. thesis, Eindhoven University of Technology. http://w3.tue.nl/en/services/library/digilib/publications_from_tue/dissertations/ (2012)
27. Van Wijk, A.C.C., Adan, I.J.B.F., Van Houtum, G.J.: Approximate evaluation of multi-location inventory models with lateral transshipments and hold back levels. Eur. J. Oper. Res. **218**, 624–635 (2012)
28. Van Wijk, A.C.C., Adan, I.J.B.F., Van Houtum, G.J.: Optimal allocation policy for a multi-location inventory system with a quick response warehouse. OR Lett. **41**, 305–310 (2013)
29. Varga, R.: Matrix Iterative Analysis. Prentice-Hall, Englewood Cliffs (1962)
30. Wong, H., Van Houtum, G.J., Cattrysse, D., Van Oudheusden, D.: Simple, efficient heuristics for multi-item multi-location spare parts systems with lateral transshipments and waiting time constraints. J. Oper. Res. Soc. **56**, 1419–1430 (2005)
31. Wong, H., Van Houtum, G.J., Cattrysse, D., Van Oudheusden, D.: Multi-item spare parts systems with lateral transshipment and waiting time constraints. Eur. J. Oper. Res. **171**, 1071–1093 (2006)
32. Xu, K., Evers, P.T., Fu, M.C.: Estimating customer service in a two-location continuous review inventory model with emergency transshipments. Eur. J. Oper. Res. **145**, 569–584 (2003)
33. Zhao, H., Deshpande, V., Ryan, J.K.: Inventory sharing and rationing in decentralized dealer networks. Manag. Sci. **51**, 531–547 (2005)
34. Zhao, H., Ryan, J.K., Deshpande, V.: Optimal dynamic production and inventory transshipment policies for a two-echelon make-to-stock system. Oper. Res. **56**, 400–410 (2008)

Chapter 6
Two-Echelon System

6.1 Introduction

In this chapter, we consider a system consisting of one central warehouse and multiple local warehouses. The local warehouses are replenished by the central warehouse. The central warehouse sends failed parts into repair at internal or external repair shops and orders new parts for consumables at external suppliers. In the core of this chapter, we assume that demands which can not be satisfied from stock at the local warehouses, are backordered; i.e., no lateral transshipments or emergency shipments are used. We consider a multi-item setting and assume basestock control. Further, we assume deterministic leadtimes for the replenishments of the local warehouses and we allow stochastic leadtimes at the central warehouse. Under these assumptions, an exact and efficient evaluation procedure can be formulated for a given basestock policy. To speed up and simplify computations, one can also use approximate evaluation procedures. We present two approximate procedures, based on two-moment fits and one-moment fits, respectively (the procedure based on the one-moment fits is the so-called METRIC approach, cf. Sherbrooke [14]).

For the optimization of basestock levels, we look at the minimization of inventory holding costs under aggregate mean waiting time constraints. We consider the case with one aggregate mean waiting time constraint per local warehouse. For this optimization problem, we describe and test a greedy heuristic and a Dantzig-Wolfe heuristic, and we show the effect of extending both heuristics with a local search. Notice that the aggregate mean waiting time constraints are equivalent to constraints for the aggregate mean numbers of backorders at the local warehouses (cf. Sect. 2.7.1).

The model assumptions that we make are the same as the common assumptions for the METRIC model of Sherbrooke [14, 15]. (We do assume deterministic replenishment leadtimes for the local warehouses, which we need for the exact evaluation procedure. Often stochastic replenishment leadtimes are assumed for the local warehouses, which we can also allow for the approximate evaluation

G.J. van Houtum, B. Kranenburg, *Spare Parts Inventory Control under System Availability Constraints*, International Series in Operations Research & Management Science 227, DOI 10.1007/978-1-4899-7609-3_6

procedures.) For the formulation of the optimization problem, we deviate from what is standard for the METRIC model. The standard is to have one overall constraint for all local warehouses together. Although this may be appropriate in some environments, OEM's or maintenance organizations generally want to realize specific targets per region. These targets may have been agreed with customers or other departments. That motivates the assumption of an aggregate mean waiting time constraint per local warehouse. The essence is that there are multiple constraints, and we deal with them in a similar way as in Chaps. 3 and 5. It is easy to adapt solution procedures for variants with constraints for subgroups of local warehouses or for multiple machine groups per local warehouse.

This chapter is organized as follows. In Sect. 6.2, we present the model. In Sects. 6.3 and 6.4, we describe the evaluation procedures and heuristics. Next, in Sect. 6.5, we compare the heuristics and we show the effect when the heuristics are based on approximate instead of exact evaluation. Finally, we conclude this chapter in Sect. 6.6.

The core of this chapter, consisting of Sects. 6.2–6.5, is largely based on Wong et al. [20].

6.2 Model

We have a non-empty set J^{loc} of local warehouses. These local warehouses are numbered $j = 1, 2, \ldots, |J^{\text{loc}}|$. Each local warehouse serves a number of installed machines of the same or at least similar type. A machine consists of several critical components, each of which may fail incidentally, and a failure of a component implies that the whole machine fails. The components are at such levels in the material breakdown structure of the machines that they can be replaced as a whole by spare parts. They are also called assemblies, and we refer to them as stock-keeping units (SKU's). Let I denote the non-empty set of all SKU's that may occur in the configurations of the machines, and the SKU's are numbered $i = 1, 2, \ldots, |I|$. We assume that the total stream of failures of SKU i as observed by local warehouse j constitutes a Poisson process with a constant rate $m_{i,j}$ (≥ 0). We define M_j as the total demand rate at local warehouse j, i.e., $M_j := \sum_{i \in I} m_{i,j}$, $j \in J^{\text{loc}}$, and we assume that $M_j > 0$.

The assumption of Poisson failure/demand processes with constant rates is standard in METRIC type models. It is a key assumption for obtaining a tractable model. For many real-life systems, lifetimes of components are (close-to) exponential, or lifetimes are not precisely exponential but the total stream of failures is a composition of sub-processes coming from relatively many machines that are supported by a local warehouse. In those cases it is reasonable to assume a Poisson failure process. Further, in practice, one does not allow long down-times of machines. Thus, it is reasonable to assume constant failure rates.

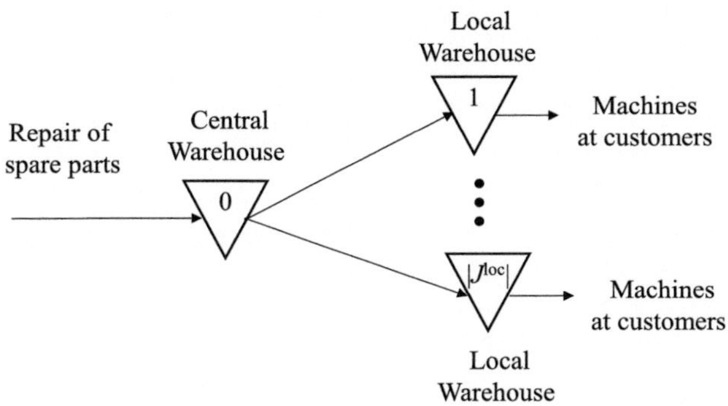

Fig. 6.1 The two-echelon system

Apart from the local warehouses, there exists a central warehouse, denoted by index 0. Let J denote the set of all warehouses, i.e., $J = \{0\} \cup J^{\text{loc}}$. A graphical representation of the two-echelon system is given in Fig. 6.1.

Suppose a part of SKU i of a machine at some local warehouse j fails. Then the machine goes down. The failed part is replaced by a spare part stocked at the local warehouse, if the local warehouse has on-hand stock. Otherwise, the demand for a spare part is backordered and the machines remain down until a spare part becomes available at the local warehouse. The failed part is shipped to the central warehouse, where all failed parts are immediately sent into repair. At the same time, a request for a spare part is placed at the central warehouse. The order and ship time for a part of SKU i from the central warehouse to local warehouse j is denoted by $t_{i,j}$. This order and ship time *excludes* a possible waiting time at the central warehouse in case a spare part is not immediately available there, and is assumed to be *deterministic*. For returned failed parts at the central warehouse, it takes a random repair leadtime with mean $t_{i,0}$ until the part is returned to the spare parts stock at the central warehouse. These repair leadtimes for parts of SKU i are assumed to be independent and identically distributed, and they are independent of the repair leadtimes for other SKU's.

In the above description, each demand for a part at a local warehouse results in an immediate replenishment order at the central warehouse. Further, at the same time a failed part is returned at the central warehouse and this failed part is immediately sent into repair. This implies that the inventory positions of a given SKU i remain at constant levels at all warehouses, or, in other words, that we have basestock control at each warehouse $j \in J$; the corresponding basestock levels are denoted by $S_{i,j}$, $j \in J$. It also implies that we have one-for-one replenishments and repairs (i.e., no batching). As we deal with critical components at the assembly level and the SKU's are expensive in general, the assumption of one-for-one replenishment is reasonable. Further, in order to ensure fairness, we assume that backordered demands for each SKU $i \in I$ at each warehouse $j \in J$ are treated in first-come first-served order. Let $\mathbf{S}_i := (S_{i,0}, S_{i,1}, \ldots, S_{i,|J^{\text{loc}}|})$, $i \in I$, denote the vector of basestock levels for SKU i, and let a basestock policy be denoted by

$$\mathbf{S} = \begin{pmatrix} S_{1,0} & S_{1,1} & \dots & S_{1,|J^{\text{loc}}|} \\ S_{2,0} & S_{2,1} & \dots & S_{2,|J^{\text{loc}}|} \\ \vdots & \vdots & \ddots & \vdots \\ S_{|I|,0} & S_{|I|,1} & \dots & S_{|I|,|J^{\text{loc}}|} \end{pmatrix}$$

As costs, we have only inventory holding costs, which are counted for all parts in the on-hand stock in either the central warehouse or one of the local warehouses. The inventory holding costs for SKU i are equal for all warehouses and they are c_i^{h} (> 0) per part on stock per unit of time. For the inventory holding costs, we exclude the parts that are in repair or in the transport pipelines from the central warehouse to the local warehouses. This has no effect on the analysis, because the inventory holding costs for the parts in repair and in transport constitute a constant factor, i.e., they are independent of the basestock policy $\mathbf{S}$. Let $C_i(\mathbf{S}_i)$ be the expected total costs per time unit for SKU i. A formula for $C_i(\mathbf{S}_i)$ will be given in Sect. 6.3. The total average costs are equal to

$$C(\mathbf{S}) = \sum_{i \in I} C_i(\mathbf{S}_i). \tag{6.1}$$

At each local warehouse j, there is a maximum level $\hat{W}_j^{\text{obj}}$ given for the aggregate mean waiting time. Under a basestock policy $\mathbf{S}$, the aggregate mean waiting time $\hat{W}_j(\mathbf{S})$ at local warehouse j is equal to

$$\hat{W}_j(\mathbf{S}) = \sum_{i \in I} \frac{m_{i,j}}{M_j} W_{i,j}(\mathbf{S}_i), \tag{6.2}$$

where $W_{i,j}(\mathbf{S}_i)$ denotes the mean waiting time for a demand for SKU i at local warehouse j.

Our goal is to minimize the total inventory holding costs subject to an aggregate mean waiting time constraint per local warehouse. This problem can be formulated as:

$$\text{(P)} \qquad \min \qquad C(\mathbf{S})$$

$$\text{subject to } \hat{W}_j(\mathbf{S}) \le \hat{W}_j^{\text{obj}}, \quad j \in J^{\text{loc}}$$

$$\mathbf{S} \in \mathscr{S},$$

where $\mathscr{S} = \{\mathbf{S} \mid S_{i,j} \in \mathbb{N}_0, \forall i \in I \text{ and } j \in J\}$. An optimal solution for Problem (P) is denoted by $\mathbf{S}^*$ and the optimal costs are denoted by C_{P}.

In the description above, we have described what happens for repairable SKU's, but the model also applies for consumable SKU's. For a consumable SKU, we assume that a new part is procured by the central warehouse each time that a demand occurs, and we assume i.i.d. procurement leadtimes with mean $t_{i,0}$ for SKU i.

6.3 Evaluation Procedures

For our model, the evaluation of a given policy can be done per SKU. The purpose of such an evaluation is to determine the costs $C_i(\mathbf{S}_i)$ and the mean waiting times $W_{i,j}(\mathbf{S}_i)$, so that the total average costs and the aggregate mean waiting times can be determined via (6.1) and (6.2). In this section we describe one exact evaluation procedure and two approximate evaluation procedures.

6.3.1 Exact Evaluation

In this section, we describe an exact evaluation procedure based on recursive expressions for the numbers of parts in repair (or on order), on-hand stocks, and numbers of backorders (cf. Graves [7]; we follow the formulae of [13], in which Graves' exact recursion has been generalized to general multi-echelon, multi-indenture systems). The evaluation of the performance can be executed per SKU.

Consider a given SKU $i \in I$ and a given basestock vector $\mathbf{S}_i$ for this SKU. We start the analysis at the central warehouse. We first need to describe the demand process for SKU i at the central warehouse. Because each demand for SKU i at a local warehouse j immediately leads to a demand for SKU i at the central warehouse, one sees also a Poisson demand process with a constant rate at the central warehouse. This constant rate, denoted by $m_{i,0}$, is equal to the sum of the rates at the local warehouses:

$$m_{i,0} = \sum_{j \in J^{\text{loc}}} m_{i,j}. \tag{6.3}$$

Each demand for SKU i at the central warehouse is accompanied by a failed part that goes into repair. Repair leadtimes for parts of SKU i are i.i.d. with mean $t_{i,0}$. Let $X_{i,0}$ be the number of parts in repair of SKU i in steady state. Then, by Palm's theorem (see [12], see also Sect. 2.3), $X_{i,0}$ is Poisson distributed with mean $m_{i,0}t_{i,0}$; i.e.,

$$P\{X_{i,0} = y\} = \frac{(m_{i,0}t_{i,0})^y}{y!} e^{-m_{i,0}t_{i,0}}, \quad y \geq 0. \tag{6.4}$$

For SKU i, the on-hand stock $OH_{i,0}$ and the number of backorders $BO_{i,0}$ at the central warehouse are equal to

$$OH_{i,0} = (S_{i,0} - X_{i,0})^+,$$
$$BO_{i,0} = (X_{i,0} - S_{i,0})^+.$$

Because they both depend on $S_{i,0}$, we denote them as $OH_{i,0}(S_{i,0})$ and $BO_{i,0}(S_{i,0})$. For their probability distributions, we obtain

$$P\{OH_{i,0}(S_{i,0}) = x\} = \begin{cases} \sum_{y=S_{i,0}}^{\infty} P\{X_{i,0} = y\}, \ x = 0, \\ P\{X_{i,0} = S_{i,0} - x\}, \ \ x \in \{1, \ldots, S_{i,0}\}, \end{cases} \tag{6.5}$$

$$P\{BO_{i,0}(S_{i,0}) = x\} = \begin{cases} \sum_{y=0}^{S_{i,0}} P\{X_{i,0} = y\}, \ x = 0, \\ P\{X_{i,0} = S_{i,0} + x\}, \ x > 0. \end{cases} \tag{6.6}$$

From these, we obtain the following expressions for the corresponding means $EOH_{i,0}(S_{i,0})$ and $EBO_{i,0}(S_{i,0})$:

$$EOH_{i,0}(S_{i,0}) = \sum_{x=0}^{S_{i,0}} (S_{i,0} - x) P\{X_{i,0} = x\}, \tag{6.7}$$

$$\begin{aligned} EBO_{i,0}(S_{i,0}) &= \sum_{x=S_{i,0}+1}^{\infty} (x - S_{i,0}) P\{X_{i,0} = x\} \\ &= m_{i,0} t_{i,0} - S_{i,0} + \sum_{x=0}^{S_{i,0}} (S_{i,0} - x) P\{X_{i,0} = x\} \\ &= m_{i,0} t_{i,0} - S_{i,0} + EOH_{i,0}(S_{i,0}). \end{aligned} \tag{6.8}$$

The next step is that we have to identify for the backordered demands at the central warehouse to which local warehouses they belong. Define $BO_{i,0}^{(j)}(S_{i,0})$, $j \in J^{\text{loc}}$, as the number of backorders of local warehouse j in the backorder queue at the central warehouse. Because each backordered demand at the central warehouse stems from local warehouse j with probability $m_{i,j}/m_{i,0}$, the probability distribution of $BO_{i,0}^{(j)}(S_{i,0})$ is obtained by

$$P\left\{BO_{i,0}^{(j)}(S_{i,0}) = x\right\} = \sum_{y=x}^{\infty} \binom{y}{x} \left(\frac{m_{i,j}}{m_{i,0}}\right)^{x} \left(1 - \frac{m_{i,j}}{m_{i,0}}\right)^{y-x} P\{BO_{i,0}(S_{i,0}) = y\}. \tag{6.9}$$

The corresponding mean $EBO_{i,0}^{(j)}(S_{i,0})$ equals

$$EBO_{i,0}^{(j)}(S_{i,0}) = \frac{m_{i,j}}{m_{i,0}} EBO_{i,0}(S_{i,0}). \tag{6.10}$$

Subsequently, we can determine the number of parts on order by local warehouse $j \in J^{\text{loc}}$. The corresponding steady-state variable is denoted by $X_{i,j}(S_{i,0})$ (it is easily verified that this variable only depends on $S_{i,0}$). This number consists of backordered orders at the central warehouse and orders for which parts are in transport from the central warehouse to local warehouse j. To derive an expression for $X_{i,j}(S_{i,0})$, we can look at the number of parts on order at an arbitrary time point t. All parts that are in transport at time $t - t_{i,j}$ will have arrived at local warehouse j at time t. The orders in the backorder queue at the central warehouse at time $t - t_{i,j}$ will arrive at the local warehouse after time t, and the same holds for the orders that are placed in the time interval $(t-t_{i,j},t]$. The latter number is equal to the number of demands in the interval $(t-t_{i,j},t]$. So,

$$\begin{aligned}X_{i,j}(S_{i,0}) &= \text{Number of parts on order by local warehouse } j \text{ at time } t\\ &= \text{Number of backorders for } j \text{ at the central warehouse at time } t-t_{i,j}\\ &\quad + \text{Number of demands at local warehouse } j \text{ in the interval } (t-t_{i,j},t]\end{aligned} \tag{6.11}$$

(notice that for this equation we use the property that the order and ship time from the central warehouse to local warehouse j is deterministic). The first term on the righthand side of this equation is equal to the number of backordered demands at the central warehouse at an arbitrary time point, and thus equal to $BO_{i,0}^{(j)}(S_{i,0})$. The second term on the righthand side is Poisson distributed with mean $m_{i,j}t_{i,j}$. We denote this Poisson distribution by $Y_{i,j}$. Hence, we can rewrite (6.11) as

$$X_{i,j}(S_{i,0}) = Y_{i,j} + BO_{i,0}^{(j)}(S_{i,0}). \tag{6.12}$$

Because $Y_{i,j}$ and $BO_{i,0}^{(j)}(S_{i,0})$ are independent, the distribution of $X_{i,j}(S_{i,0})$ can be obtained by convoluting the distributions of $Y_{i,j}$ and $BO_{i,0}^{(j)}(S_{i,0})$.

Via $X_{i,j}(S_{i,0})$ and by exploiting the fact that the inventory at local warehouse j is controlled by a basestock policy with basestock level $S_{i,j}$, we obtain the on-hand stock $OH_{i,j}(S_{i,0},S_{i,j})$ and number of backorders $BO_{i,j}(S_{i,0},S_{i,j})$ at local warehouse j by the following equations:

$$\begin{aligned}OH_{i,j}(S_{i,0},S_{i,j}) &= (S_{i,j} - X_{i,j}(S_{i,0}))^+,\\ BO_{i,j}(S_{i,0},S_{i,j}) &= (X_{i,j}(S_{i,0}) - S_{i,j})^+.\end{aligned}$$

For their probability distributions, we obtain

$$P\{OH_{i,j}(S_{i,0},S_{i,j}) = x\} = \begin{cases}\sum_{y=S_{i,j}}^{\infty} P\{X_{i,j}(S_{i,0}) = y\}, \; x=0,\\ P\{X_{i,j}(S_{i,0}) = S_{i,j} - x\}, \quad x \in \{1,\dots,S_{i,j}\},\end{cases} \tag{6.13}$$

$$P\{BO_{i,j}(S_{i,0},S_{i,j}) = x\} = \begin{cases}\sum_{y=0}^{S_{i,j}} P\{X_{i,j}(S_{i,0}) = y\}, \; x=0,\\ P\{X_{i,j}(S_{i,0}) = S_{i,j} + x\}, \; x>0.\end{cases} \tag{6.14}$$

And the corresponding means $EOH_{i,j}(S_{i,0},S_{i,j})$ and $EBO_{i,j}(S_{i,0},S_{i,j})$ are equal to (similar to (6.7)–(6.8))

$$EOH_{i,j}(S_{i,0},S_{i,j}) = \sum_{x=0}^{S_{i,j}} (S_{i,j} - x) P\{X_{i,j}(S_{i,0}) = x\}, \tag{6.15}$$

$$
\begin{aligned}
EBO_{i,j}(S_{i,0},S_{i,j}) &= \sum_{x=S_{i,j}+1}^{\infty} (x-S_{i,j})P\{X_{i,j}(S_{i,0})=x\} \\
&= E\left\{X_{i,j}(S_{i,0})\right\} - S_{i,j} + \sum_{x=0}^{S_{i,j}} (S_{i,j}-x)P\{X_{i,j}(S_{i,0})=x\} \\
&= m_{i,j}t_{i,j} + EBO_{i,0}^{(j)}(S_{i,0}) - S_{i,j} + EOH_{i,j}(S_{i,0},S_{i,j}). \qquad (6.16)
\end{aligned}
$$

Finally, by applying Little's law to the backordered demands at local warehouse j, we find

$$
W_{i,j}(\mathbf{S}_i) = \frac{EBO_{i,j}(S_{i,0},S_{i,j})}{m_{i,j}}. \qquad (6.17)
$$

Notice that $W_{i,j}(\mathbf{S}_i)$ depends on $S_{i,0}$ and $S_{i,j}$ only. Further, we may use the expressions for $EOH_{i,0}(S_{i,0})$ and the $EOH_{i,j}(S_{i,0},S_{i,j})$, $j \in J^{\text{loc}}$, to determine the costs $C_i(\mathbf{S}_i)$ for SKU i by the following formula:

$$
C_i(\mathbf{S}_i) = c_i^{\text{h}} \left(EOH_{i,0}(S_{i,0}) - \sum_{j \in J^{\text{loc}}} EOH_{i,j}(S_{i,0},S_{i,j}) \right). \qquad (6.18)
$$

The formal description of the exact evaluation for a given SKU i and a given basestock vector $\mathbf{S}_i$ is described in Algorithm 6.1.

Algorithm 6.1 *(Exact evaluation procedure)*

Step 1 Determine $m_{i,0}$ by (6.3);
Determine the distribution of $X_{i,0}$ by (6.4);
Determine the distribution of $BO_{i,0}(S_{i,0})$ by (6.6)
Step 2 For all $j \in J^{\text{loc}}$:
Determine the distribution of $BO_{i,0}^{(j)}(S_{i,0})$ by (6.9);
Determine the distribution of $X_{i,j}(S_{i,0})$ by (6.12);
Determine $EBO_{i,j}(S_{i,0},S_{i,j})$ by (6.16)
Step 3 Determine the $W_{i,j}(\mathbf{S}_i)$, $j \in J^{\text{loc}}$, by (6.17);
Determine $C_i(\mathbf{S}_i)$ by (6.18)

When executing the above algorithm, for the random variables with support $\mathbb{N}_0$, the calculation of the probability distributions can be stopped at a sufficiently large value. For a random variable Y on support $\mathbb{N}_0$, we compute the probabilities $P\{Y = y\}$ for $y = 0, \ldots, y^{\text{trunc}}$, where $y^{\text{trunc}} = \min\{y | P\{Y \leq y\} \geq 1-\varepsilon\}$ and $\varepsilon = 10^{-6}$, and we allocate the remaining probability mass $1 - P\{Y \leq y^{\text{trunc}}\}$ to $P\{Y = y^{\text{trunc}}\}$.

In Algorithm 6.1, a large part of the computational effort is spent to the computation of the whole distributions of the various random variables and the convolutions to determine the $X_{i,j}(S_{i,0})$. This can be avoided in the approximate evaluation procedures.

Example 6.1. We consider a system with $|J^{\mathrm{loc}}| = 2$ local warehouses and one SKU i with:

$$\begin{aligned} &m_{i,1} = 0.1 \text{ demands/week},\ \ m_{i,2} = 0.2 \text{ demands/week},\\ &t_{i,0} = 4 \text{ weeks},\ \ t_{i,1} = t_{i,2} = 1 \text{ week},\\ &S_{i,0} = 2,\ \ S_{i,1} = S_{i,2} = 1. \end{aligned}$$

Then the demand rate of this SKU i at the central warehouse is $m_{i,0} = 0.3$ demands/week. We apply Algorithm 6.1 to determine $W_{i,1}(\mathbf{S}_i)$. Via Step 1, we find the following values for the backorder probabilities $P\{BO_{i,0}(S_{i,0}) = x\}, x = 0,\ldots,6$: 0.8795, 0.0867, 0.0260, 0.0063, 0.0013, 0.0002, 0.0000. Subsequently, in Step 2, for $P\{BO_{i,1}(S_{i,0}, S_{i,1}) = x\}$, $x = 0,\ldots,3$, we obtain: 0.9864, 0.0126, 0.0010, 0.0000. And the mean number of backorders at local warehouse 1 is equal to $EBO_{i,1}(S_{i,0}, S_{i,1}) = 0.0146$. Finally, in Step 3, we find that $W_{i,1}(\mathbf{S}_i) = 0.146$ weeks. In Problem 6.1, it is asked to determine $W_{i,2}(\mathbf{S}_i)$.

6.3.2 Approximate Evaluation Based on Two-Moment Fits

In the first approximate evaluation procedure, we switch from computing whole distributions to computing the first two moments of the various random variables. With "the first two moments", we refer to the mean and either the second moment or the variance. Notice that if one has determined the second moment of a random variable, in addition to the mean, then we know also the variance, and the other way around. The approximate evaluation method based on two moments stems from Graves [7]; we follow again the formulae of [13].

Consider a given SKU $i \in I$ and a given basestock vector $\mathbf{S}_i$ for this SKU. Let us follow the order in which random variables are determined in Algorithm 6.1, and denote how the first two moments can be obtained.

The number of parts in repair at the central warehouse, $X_{i,0}$, is Poisson distributed with mean $m_{i,0}t_{i,0}$, and thus (cf. (6.35)–(6.36))

$$E\{X_{i,0}\} = Var\{X_{i,0}\} = m_{i,0}t_{i,0},\ \ E\{X_{i,0}^2\} = m_{i,0}t_{i,0}(m_{i,0}t_{i,0} + 1). \tag{6.19}$$

Next, the first moment of $BO_{i,0}(S_{i,0})$ is given by (6.8). For the second moment and the variance, we obtain

$$\begin{aligned} E\{(BO_{i,0}(S_{i,0}))^2\} &= \sum_{x=S_{i,0}+1}^{\infty} (x - S_{i,0})^2 P\{X_{i,0} = x\}\\ &= \sum_{x=0}^{\infty} (S_{i,0} - x)^2 P\{X_{i,0} = x\} - \sum_{x=0}^{S_{i,0}} (S_{i,0} - x)^2 P\{X_{i,0} = x\} \end{aligned}$$

$$= E\{X_{i,0}^2\} - 2S_{i,0}E\{X_{i,0}\} + S_{i,0}^2 - \sum_{x=0}^{S_{i,0}} (S_{i,0} - x)^2 P\{X_{i,0} = x\}, \quad (6.20)$$

$$Var\{BO_{i,0}(S_{i,0})\} = E\{(BO_{i,0}(S_{i,0}))^2\} - (EBO_{i,0}(S_{i,0}))^2. \quad (6.21)$$

Subsequently, we need to determine the first two moments of $BO_{i,0}^{(j)}(S_{i,0})$ for each $j \in J^{\text{loc}}$. The first moment is given by (6.10). For the second moment, we use the property that, if $BO_{i,0}(S_{i,0}) = y$, then $BO_{i,0}^{(j)}(S_{i,0})$ is binomially distributed with y trials and success probability $f_{i,j} := m_{i,j}/m_{i,0}$. We find (use also (6.32))

$$\begin{aligned}
E\{(BO_{i,0}^{(j)}(S_{i,0}))^2\} &= \sum_{x=0}^{\infty} E\{(BO_{i,0}^{(j)}(S_{i,0}))^2 | BO_{i,0}(S_{i,0}) = x\} P\{BO_{i,0}(S_{i,0}) = x\} \\
&= \sum_{x=0}^{\infty} x f_{i,j}[(x-1)f_{i,j} + 1] P\{BO_{i,0}(S_{i,0}) = x\} \\
&= \sum_{x=0}^{\infty} [f_{i,j}(1 - f_{i,j})x + f_{i,j}^2 x^2] P\{BO_{i,0}(S_{i,0}) = x\} \\
&= f_{i,j}(1 - f_{i,j}) EBO_{i,0}(S_{i,0}) + f_{i,j}^2 E\{(BO_{i,0}(S_{i,0}))^2\}. \quad (6.22)
\end{aligned}$$

The variance is obtained via the first two moments:

$$\begin{aligned}
Var\{BO_{i,0}^{(j)}(S_{i,0})\} &= E\{(BO_{i,0}^{(j)}(S_{i,0}))^2\} - (EBO_{i,0}^{(j)}(S_{i,0}))^2 \\
&= f_{i,j}^2 Var\{BO_{i,0}(S_{i,0})\} + f_{i,j}(1 - f_{i,j}) EBO_{i,0}(S_{i,0}). \quad (6.23)
\end{aligned}$$

Then, by (6.11) and using the property that $Y_{i,j}$ is Poisson distributed with mean $m_{i,j}t_{i,j}$, we obtain the first two moments and variance of $X_{i,j}(S_{i,0})$:

$$E\{X_{i,j}(S_{i,0})\} = m_{i,j}t_{i,j} + E\{BO_{i,0}^{(j)}(S_{i,0})\}, \quad (6.24)$$

$$Var\{X_{i,j}(S_{i,0})\} = m_{i,j}t_{i,j} + Var\{BO_{i,0}^{(j)}(S_{i,0})\}, \quad (6.25)$$

$$E\{(X_{i,j}(S_{i,0}))^2\} = Var\{X_{i,j}(S_{i,0})\} + (E\{X_{i,j}(S_{i,0})\})^2. \quad (6.26)$$

At this point, we apply a so-called two-moment fit. This is needed to obtain the probabilities $P\{X_{i,j}(S_{i,0}) = x\}$, $x = 0, \ldots, S_{i,j}$, which are needed to compute $EOH_{i,j}(S_{i,0}, S_{i,j})$ and $EBO_{i,j}(S_{i,0}, S_{i,j})$ by (6.15) and (6.16). Applying a two-moment fit means that a given probability distribution is taken and that its parameters are chosen such that the resulting distribution has first two moments that are equal to $E\{X_{i,j}(S_{i,0})\}$ and $E\{(X_{i,j}(S_{i,0}))^2\}$. In Appendix "Fitting Discrete Distributions on the First Two Moments", two methods are discussed for this two-moment fit:

- A method that fits a negative binomial distribution or a Poisson distribution. For this method, it is required that $Var\{X_{i,j}(S_{i,0})\} \geq E\{X_{i,j}(S_{i,0})\}$, i.e., that the variance-to-mean ratio of $X_{i,j}(S_{i,0})$ is at least 1. If it is 1, then one fits a Poisson distribution with $\lambda = E\{X_{i,j}(S_{i,0})\}$. If it is strictly larger than 1, then one can fit a negative binomial distribution method as described in Appendix "Two-Moment Fit by a Negative Binomial Distribution". Graves [7] used this fitting method; he established by a computational experiment that one always has a variance-to-mean ratio of $X_{i,j}(S_{i,0})$ of at least 1 (see [7], p. 1252).

- The method of Adan et al. [1], as described in Appendix "Two-Moment Fit of Adan et al. [1]". In this method, multiple distributions are used and it depends on the values of $Var\{X_{i,j}(S_{i,0})\}$ and $E\{X_{i,j}(S_{i,0})\}$ which distribution is used. Under this method, it is not required that the variance-to-mean ratio of $X_{i,j}(S_{i,0})$ is at least 1.

The use of this two-moment fit is the only approximate step in the whole evaluation method as described in this chapter. Because two moments are used for this fit, the accuracy is generally very good. This holds for both fitting methods. So, one can choose one of these methods.

Once the two-moment fit has been applied, one can execute the rest of Algorithm 6.1 as before. This leads to the following approximate evaluation method.

Algorithm 6.2 *(Approximate evaluation based on two-moment fits)*

Step 1 Determine $m_{i,0}$ by (6.3);
Determine the first two moments and variance of $X_{i,0}$ by (6.19);
Determine the first two moments and variance of $BO_{i,0}(S_{i,0})$ by (6.8), (6.20), (6.21)
Step 2 For all $j \in J^{\text{loc}}$:
Determine the first two moments and variance of $BO_{i,0}^{(j)}(S_{i,0})$ by (6.10), (6.22), (6.23);
Determine the first two moments and variance of $X_{i,j}(S_{i,0})$ by (6.24), (6.25), (6.26);
Apply a two-moment fit for $X_{i,j}(S_{i,0})$ (cf. Appendix "Two-Moment Fit by a Negative Binomial Distribution" or Appendix "Two-Moment Fit of Adan et al. [1]");
Determine $EBO_{i,j}(S_{i,0},S_{i,j})$ by (6.16)
Step 3 Determine the $W_{i,j}(\mathbf{S}_i)$, $j \in J^{\text{loc}}$, by (6.17);
Determine $C_i(\mathbf{S}_i)$ by (6.18)

Example 6.1 (continued). We apply Algorithm 6.2 to determine $W_{i,1}(\mathbf{S}_i)$, where we use a negative binomial distribution for the two-moment fit for $X_{i,1}(S_{i,0})$ (cf. Appendix "Two-Moment Fit by a Negative Binomial Distribution"). In Step 1, we find $EBO_{i,0}(S_{i,0}) = 0.1638$, $E\{(BO_{i,0}(S_{i,0}))^2\} = 0.2738$, and $Var\{BO_{i,0}(S_{i,0})\} = 0.2470$. In Step 2, we find $E\{X_{i,1}(S_{i,0})\} = 0.1546$, $Var\{X_{i,1}(S_{i,0})\} = 0.1638$, and $E\{(X_{i,1}(S_{i,0}))^2\} = 0.1877$. For the two-moment fit on $X_{i,0}(S_{i,0})$, we obtain a negative binomial distribution with $p = 0.05617$ and $k = 2.598$. Next, the mean number of backorders at local warehouse 1 is approximated by $EBO_{i,1}(S_{i,0},S_{i,1}) = 0.0152$. Finally, as approximation for $W_{i,1}(\mathbf{S}_{i,1})$, we obtain 0.152 weeks. This is 0.006 weeks higher than the exact value.

6.3.3 METRIC Approach

The approximate evaluation procedure of the previous section can be further simplified by going back to single-moment calculations for the various random variables, and by simply approximating the distribution of $X_{i,j}(S_{i,0})$, $j \in J^{\text{loc}}$, by a Poisson distribution with $\lambda = E\{X_{i,j}(S_{i,0})\}$. The latter fit is a single-moment fit, and, obviously, such a fit can lead to a much larger inaccuracy than two-moment fits. The resulting approximate evaluation method is the called the *METRIC approach*, as proposed by Sherbrooke [14] in 1968. The procedure is described in more detail in Algorithm 6.3.

Algorithm 6.3 *(METRIC approach: Approximate evaluation via single-moment fits)*

Step 1 Determine $m_{i,0}$ by (6.3);
 Determine the mean of $BO_{i,0}(S_{i,0})$ by (6.8)
Step 2 For all $j \in J^{\text{loc}}$:
 Determine the mean of $BO_{i,0}^{(j)}(S_{i,0})$ by (6.10);
 Determine the mean of $X_{i,j}(S_{i,0})$ by (6.24);
 Approximate the distribution of $X_{i,j}(S_{i,0})$ by a Poisson distribution with $\lambda = E\{X_{i,j}(S_{i,0})\}$;
 Determine $EBO_{i,j}(S_{i,0}, S_{i,j})$ by (6.16)
Step 3 Determine the $W_{i,j}(\mathbf{S}_i)$, $j \in J^{\text{loc}}$, by (6.17);
 Determine $C_i(\mathbf{S}_i)$ by (6.18)

The only approximate step in this approximation procedure is the fit of a Poisson distribution on the random variables $X_{i,j}(S_{i,0})$. In Sect. 6.3.2, we mentioned already an observation of Graves [7]. He established by a computational experiment that one always has a variance-to-mean ratio of $X_{i,j}(S_{i,0})$ of at least 1 (see [7], p. 1252). A Poisson distribution has a variance-to-mean ratio of 1, and thus a fit of a Poisson distribution on the a random variable $X_{i,j}(S_{i,0})$ implies that the approximated distribution is less variable than the exact distribution. If the basestock level $S_{i,j}$ for is large, this will lead to an underestimation of $EBO_{i,j}(S_{i,0}, S_{i,j})$ and $W_{i,j}(\mathbf{S}_i)$ at local warehouse j. If the basestock level $S_{i,j}$ is small, this will lead to an overestimation of $EBO_{i,j}(S_{i,0}, S_{i,j})$ and $W_{i,j}(\mathbf{S}_i)$.

Example 6.1 (continued). Let us apply the METRIC approach to approximate $W_{i,1}(\mathbf{S}_{i,1})$. In Step 1, we find $EBO_{i,0}(S_{i,0}) = 0.163$. Subsequently, in Step 2, we obtain $EBO_{i,0}^{(1)}(S_{i,0}) = 0.0546$ and $EX_{i,1}(S_{i,0}) = 0.1546$. For $P\{BO_{i,1}(S_{i,0}, S_{i,1}) = x\}$, $x = 0, \ldots, 3$, we obtain: 0.9893, 0.0102, 0.0005, 0.0000. As approximation for the mean number of backorders at local warehouse 1, we get $EBO_{i,1}(S_{i,0}, S_{i,j}) = 0.0112$. Finally, as approximation for $W_{i,1}(\mathbf{S}_{i,1})$, we obtain 0.121 weeks. This is 0.034 weeks lower than the exact value.

6.4 Heuristics

In this section, we describe heuristic methods for solving Problem (P). In Sect. 6.4.1, we describe a greedy heuristic. Next, in Sect. 6.4.2, we describe a local search procedure by which a greedy solution can be further improved. Subsequently, in Sect. 6.4.3, we apply Dantzig-Wolfe decomposition, which leads to both a heuristic solution and a lower bound for the optimal costs of Problem (P). Also the solution of the Dantzig-Wolfe heuristic can be further improved by a local search. This leads to 4 heuristic methods in total; see Sect. 6.4.4.

6.4.1 Greedy Heuristic

Our greedy heuristic is similar to the greedy heuristics for the problems of Chaps. 3 and 5; see Algorithms 3.1 and 5.3 Here the costs $C(\mathbf{S})$ consist only of inventory holding costs. This implies that they are minimized by setting all basestock levels equal to 0 and that next one can immediately start with greedy steps, in which one increases the basestock level $S_{i,j}$ leading to the largest decrease in distance to the set of feasible solutions per unit increase in costs.

The greedy heuristic is formally described as follows. First of all, we divide the set $\mathscr{S}$ of all solutions into a subset

$$\mathscr{S}^{\text{feas}} := \{\mathbf{S} \in \mathscr{S} | \hat{W}_j(\mathbf{S}) \leq \hat{W}_j^{\text{obj}} \text{ for all } j \in J^{\text{loc}}\}$$

of feasible solutions and the subset $\mathscr{S} \setminus \mathscr{S}^{\text{feas}}$ of non-feasible solutions. Next, for each solution $\mathbf{S} \in \mathscr{S}$, we define the distance $d(\mathbf{S})$ to the set $\mathscr{S}^{\text{feas}}$ of feasible solutions as

$$\begin{aligned} d(\mathbf{S}) &:= \sum_{j \in J^{\text{loc}}} \left(\hat{W}_j(\mathbf{S}) - \hat{W}_j^{\text{obj}} \right)^+ \\ &= \sum_{j \in J^{\text{loc}}} \left(\sum_{i \in I} \frac{m_{i,j}}{M_j} W_{i,j}(\mathbf{S}_i) - \hat{W}_j^{\text{obj}} \right)^+ . \end{aligned} \tag{6.27}$$

In each greedy step, we proceed as follows. We have a current solution $\mathbf{S}$, and we look at the ratio of the decrease in distance to the set of feasible solutions and the increase in costs if one of the basestock levels $S_{i,j}$ is increased by one unit. Let $i \in I$ and $j \in J$, and suppose that $S_{i,j}$ would be increased by 1 unit. Then the increase in costs is equal to

$$\Delta_{i,j} C(\mathbf{S}) = \Delta_j C_i(\mathbf{S}_i) = C_i(\mathbf{S}_i + \mathbf{e}_j) - C_i(\mathbf{S}_i), \tag{6.28}$$

where $\mathbf{e}_j$ is a $|J|$-dimensional vector with a 1 on the j-th position (the positions are numbered as $0, 1, \ldots, |J^{\text{loc}})$ and zeros on all other positions.

The decrease in distance to the feasible solutions is given by $-\Delta_{i,j}d(\mathbf{S})$. For each $i \in I$ and $j \in J$, let $\mathbf{E}_{i,j}$ be an $|I| \times |J|$ matrix with positions (i', j'), $i' \in I$, $j' \in J$, a 1 on the (i, j)-th position and zeros on all other positions. For $j = 0$, we obtain

$$\begin{aligned}
\Delta_{i,0}d(\mathbf{S}) &= d(\mathbf{S}+\mathbf{E}_{i,0}) - d(\mathbf{S}) \\
&= \sum_{j' \in J^{\text{loc}}} \left[\left(\hat{W}_{j'}(\mathbf{S}+\mathbf{E}_{i,0}) - \hat{W}_{j'}^{\text{obj}} \right)^+ - \left(\hat{W}_{j'}(\mathbf{S}) - \hat{W}_{j'}^{\text{obj}} \right)^+ \right] \\
&= \sum_{j' \in J^{\text{loc}}} \left[\left(\sum_{i' \in I \setminus \{i\}} \frac{m_{i',j'}}{M_{j'}} W_{i',j'}(\mathbf{S}_{i'}) + \frac{m_{i,j'}}{M_{j'}} W_{i,j'}(\mathbf{S}_i + \mathbf{e}_0) - \hat{W}_{j'}^{\text{obj}} \right)^+ \right. \\
&\qquad \left. - \left(\sum_{i' \in I} \frac{m_{i',j'}}{M_{j'}} W_{i',j'}(\mathbf{S}_{i'}) - \hat{W}_{j'}^{\text{obj}} \right)^+ \right]. \qquad (6.29)
\end{aligned}$$

For all other j, we can exploit the property that increasing $S_{i,j}$ has no effect on the mean waiting times $W_{i,j'}(\mathbf{S})$ for all $j' \in J^{\text{loc}} \setminus \{j\}$ (see the remark right after (6.17)):

$$\begin{aligned}
\Delta_{i,j}d(\mathbf{S}) &= d(\mathbf{S}+\mathbf{E}_{i,j}) - d(\mathbf{S}) \\
&= \sum_{j' \in J^{\text{loc}}} \left[\left(\hat{W}_{j'}(\mathbf{S}+\mathbf{E}_{i,j}) - \hat{W}_{j'}^{\text{obj}} \right)^+ - \left(\hat{W}_{j'}(\mathbf{S}) - \hat{W}_{j'}^{\text{obj}} \right)^+ \right] \\
&= \left(\hat{W}_{j}(\mathbf{S}+\mathbf{E}_{i,j}) - \hat{W}_{j}^{\text{obj}} \right)^+ - \left(\hat{W}_{j}(\mathbf{S}) - \hat{W}_{j}^{\text{obj}} \right)^+ \\
&= \left(\sum_{i' \in I \setminus \{i\}} \frac{m_{i',j}}{M_{j}} W_{i',j}(\mathbf{S}_{i'}) + \frac{m_{i,j}}{M_{j}} W_{i,j}(\mathbf{S}_i + \mathbf{e}_j) - \hat{W}_{j}^{\text{obj}} \right)^+ \\
&\qquad - \left(\sum_{i' \in I} \frac{m_{i',j}}{M_{j}} W_{i',j}(\mathbf{S}_{i'}) - \hat{W}_{j}^{\text{obj}} \right)^+, \quad j \in J^{\text{loc}}. \qquad (6.30)
\end{aligned}$$

It may be shown that increasing the basestock level $S_{i,0}$ for an SKU $i \in I$ at the central warehouse has a decreasing effect on the mean waiting times $W_{i,j}(\mathbf{S}_i)$ at all local warehouses $j \in J^{\text{loc}}$, and no effect on the mean waiting times for other SKU's. Further, increasing the basestock level $S_{i,j}$ for an SKU $i \in I$ at a local warehouse $j \in J^{\text{loc}}$ has a decreasing effect on the mean waiting times $W_{i,j}(\mathbf{S}_i)$ for that SKU at the same local warehouse, and no effect on all other mean waiting times. This may be shown via the exact evaluation procedure of Sect. 6.3.1 (one can show that certain random variables for numbers of backorders become stochastically smaller; see also Problem 6.4). By (6.29) and (6.30), it is then easily verified that $\Delta_{i,j}d(\mathbf{S}) \geq 0$ for all $i \in I$ and $j \in J$. The factor

$$\Gamma_{i,j} = \frac{-\Delta_{i,j}d(\mathbf{S})}{\Delta_{i,j}C(\mathbf{S})} \qquad (6.31)$$

denotes the decrease in distance to the feasible solutions per unit of increase in costs. For the combination of SKU i and local warehouse j with the highest ratio $\Gamma_{i,j}$, we have the "biggest bang for the buck" and we increase the basestock level to $S_{i,j}+1$. We continue with the above greedy steps until we arrive at a feasible solution. The formal procedure is described in Algorithm 6.4.

Algorithm 6.4 (Greedy heuristic)

Step 1 Set $S_{i,j} := 0,\ i \in I,\ j \in J$
Step 2

Step 2-a Calculate $\Delta_{i,j} d(\mathbf{S})$, and $\Gamma_{i,j},\ i \in I,\ j \in J$ (cf. (6.29)–(6.31))
Step 2-b While $d(\mathbf{S}) > 0$:
(a) Determine i' and j' such that $\Gamma_{i',j'} \geq \Gamma_{i,j},\ i \in I,\ j \in J$;
(b) Set $S_{i',j'} := S_{i',j'} + 1$;
(c) Calculate $\Delta_{i,j} d(\mathbf{S})$, and $\Gamma_{i,j},\ i \in I,\ j \in J$ (cf. (6.29)–(6.31))

Computation time can be saved in the execution of the algorithm if only those (intermediate) results are updated that are affected by a certain change.

6.4.2 Local Search

Once a feasible solution has been obtained, one may apply a local search method to obtain a further improved solution. We apply a greedy (steepest descent) local search method that allows us to explore the entire neighborhood. At each iteration, all possible neighbors of the current solution are evaluated, and the one with the lowest total costs is selected. If the new total costs are strictly lower than the current total costs, the selected solution becomes the new solution, and the local search is further continued. Otherwise, no local improvement is possible and we take the current solution as the heuristic solution.

For each solution $\mathbf{S}$, we define the neighborhood of $\mathbf{S}$, $\mathscr{N}(\mathbf{S})$, as follows:

$$\mathscr{N}(\mathbf{S}) := \mathscr{N}_1(\mathbf{S}) \cup \mathscr{N}_2(\mathbf{S}) \cup \mathscr{N}_3(\mathbf{S}) \cup \mathscr{N}_4(\mathbf{S}),$$

with

$$\mathscr{N}_1(\mathbf{S}) := \left\{ \mathbf{S}' \in \mathscr{S}^{\text{feas}} | \mathbf{S}' = \mathbf{S} - \mathbf{E}_{i,j}, i \in I, j \in J \right\},$$
$$\mathscr{N}_2(\mathbf{S}) := \left\{ \mathbf{S}' \in \mathscr{S}^{\text{feas}} | \mathbf{S}' = \mathbf{S} + \mathbf{E}_{i,j}, i \in I, j \in J \right\},$$
$$\mathscr{N}_3(\mathbf{S}) := \left\{ \mathbf{S}' \in \mathscr{S}^{\text{feas}} | \mathbf{S}' = \mathbf{S} + \mathbf{E}_{i,j} - \mathbf{E}_{i',j'}, i \in I, i' \in I, i \neq i', j \in J, j' \in J \right\},$$
$$\mathscr{N}_4(\mathbf{S}) := \left\{ \mathbf{S}' \in \mathscr{S}^{\text{feas}} | \mathbf{S}' = \mathbf{S} + \mathbf{E}_{i,j} - \mathbf{E}_{i,j'}, i \in I, j \in J, j' \in J, j \neq j' \right\}.$$

The neighborhood of a solution is the union of four sub-neighborhoods. The first sub-neighborhood is formed by reducing one unit of stock in the system. Obviously, any possible move to the first sub-neighborhood would always give a cheaper solution. The second sub-neighborhood is formed by adding one unit of stock, which always leads to a more expensive policy. Hence, the solutions of this second sub-neighborhood are never selected in the local search procedure. However, we still include this sub-neighborhood because we use the increase in costs when moving to these neighbors for the calculations for the third sub-neighborhood; see below. The third sub-neighborhood is formed by removing one unit of an SKU and putting an extra unit of another SKU on stock, either at the same or at another location. A costs reduction may be obtained here when a unit of an expensive SKU is replaced by a unit of a less expensive SKU. Lastly, exploring the fourth sub-neighborhood may be useful to obtain the best stock allocation across all warehouses for each SKU. Both the first and second sub-neighborhood contains at most $|I||J|$ elements, the third sub-neighborhood has at most $|I|(|I|-1)(|J|)^2$ solutions, and the fourth sub-neighborhood has at most $|I||J|(|J|-1)$ solutions. Thus, the upper bound on the neighborhood size of a solution is equal to $|I||J|(2+(|I|-1)|J|+|J|-1) = |I||J|(1+|I||J|)$.

At the first iteration, we need to evaluate at most $2|I||J|$ neighboring solutions for the first and second sub-neighborhoods, and $|I||J|(|J|-1)$ solutions for the fourth sub-neighborhood. To evaluate a solution lying in the third sub-neighborhood, we can use the results obtained from the first and second sub-neighborhood. At the subsequent iterations, we only need to evaluate one or a few new neighbors since any changes of the stock levels for a given SKU do not affect the results for the other SKU's.

6.4.3 Dantzig-Wolfe Decomposition

In this section, we describe the application of Dantzig-Wolfe decomposition [6], which leads to another heuristic solution and a lower bound for the optimal costs C_P of Problem (P).

We first introduce the *Master Problem*, in which the variables of our original problem (basestock levels) are expressed as convex combinations of columns that contain possible values for the decision variables in the original problem. Besides the Master Problem, a *Restricted Master Problem* is defined that only considers a subset of all possible columns. The method starts with some initial columns which constitute a feasible solution for the Restricted Master Problem, and solves the Restricted Master Problem with the simplex method. Next, the method iteratively solves a *Sub-Problem* for each SKU $i \in I$, to determine if there exists a column for that SKU that would improve the solution, adds this column to the Restricted Master Problem, and solves the Restricted Master Problem.

The Sub-Problem for one SKU i is

$$\min \qquad c_i^{\mathrm{h}} EOH_{i,0}(S_{i,0}) + \sum_{j \in J^{\mathrm{loc}}} c_i^{\mathrm{h}} EHO_{i,j}(S_{i,0}, S_{i,j}) - \sum_{j \in J^{\mathrm{loc}}} u_j \frac{EBO_{i,j}(S_{i,0}, S_{i,j})}{\sum_{k \in I} m_{k,j}} - v_i$$

$$\text{subject to } S_{i,j} \in \mathbb{N}_0, \quad i \in I, \;\; j \in J,$$

where u_j is the dual price of the waiting time constraint for local warehouse $j \in J^{\mathrm{loc}}$ in the Restricted Master Problem, and v_i is the dual price for a constraint in the Restricted Master Problem that assures that for SKU i a convex combination of policies is chosen. Dual price $u_j \leq 0$ by definition for all $j \in J^{\mathrm{loc}}$. This Sub-Problem comes down to solving a single-item costs minimization problem with linear inventory and backordering costs (but without service level constraints). This optimization problem is *precisely the problem studied by Axsäter [2]* and we solve this problem by using his method. If the resulting policy for an SKU i has negative reduced costs (i.e., a negative value of the objective function of the Sub-Problem), this policy is added as a column to the Restricted Master Problem. The method ends if for none of the SKU's a policy with negative reduced costs is found.

Via the above procedure, we obtain an optimal solution for the Master Problem. In general, the corresponding policy will not be a basestock policy, but a convex combination of basestock policies (columns in the Restricted Master Problem), which constitutes a randomized policy. The optimal costs are denoted by C_{MP}, and C_{MP} is a lower bound for C_{P}.

Next, we obtain a heuristic solution for Problem (P) as follows. We first construct a starting policy: For each $i \in I$ and $j \in J$, we select the smallest value for $S_{i,j}$ that is found among the basestock levels of the convex combination. This leads to a solution $\mathbf{S}$ that is not feasible in general. Next, we apply greedy steps, as in Step 2 of the greedy algorithm, to obtain a feasible solution. We refer to the overall procedure as the Dantzig-Wolfe heuristic.

Like the greedy heuristic, also the solution of the Dantzig-Wolfe heuristic can be further improved by the local search procedure of Sect. 6.4.2.

6.4.4 Overview of Heuristics

In the previous sections, we have described the greedy heuristic and the Dantzig-Wolfe heuristic. For both, one has the option to apply local search to further improve the generated solution. This leads in total to the following 4 heuristics (in parentheses, we denote abbreviations):

Heuristic 1: Greedy heuristic (Greedy)
Heuristic 2: Greedy heuristic + Local search (Greedy + LS)
Heuristic 3: Dantzig-Wolfe heuristic (DW)
Heuristic 4: Dantzig-Wolfe heuristic + Local search (DW + LS)

For each heuristic, we get an upper bound for the optimality gap via the lower bound C_{MP} obtained by Dantzig-Wolfe decomposition.

6.5 Computational Results

In this section, we execute computational experiments to compare the four heuristics on the basis of their gaps with the Dantzig-Wolfe lower bound and their computation times. The test beds for the experiments are described in Sect. 6.5.1, and the results of the comparison are presented in Sect. 6.5.2. For this comparison, we use exact evaluations in all heuristics, so that fair conclusions on the heuristics itself can be drawn. For large problem instances as they occur in practice, it is desired to speed up computations, which may be done by using approximate evaluations. Therefore, in Sect. 6.5.3, we show the effect of the use of approximate evaluations.

6.5.1 Setup of Test Beds

Four test beds are considered in our experiment. In the first test bed, we consider symmetric cases in which the demand rates across all local warehouses are identical, but they are varied for different SKU's. In the second test bed, we consider asymmetric cases in which the demand rates for different local warehouses are different. For those two test beds, the target aggregate mean waiting times of all local warehouses are identical. To see how the heuristics perform when the local warehouses have different targets, we introduce a third and fourth test bed.

Test bed 1: For this test bed, we construct instances that are symmetric across all local warehouses. We consider two different numbers of local warehouses ($|J^{\text{loc}}| =$ 5 and 20) and two different numbers of SKU's ($|I|$ = 20 and 100). With regard to the demand rates $m_{i,j}$, for each $i \in I$, we draw $m_{i,1}$ from the uniform distribution U[0.002, 0.08] and we take $m_{i,j} = m_{i,1}$ for all other $j \in J^{\text{loc}}$. Similarly, symmetric values for the inventory holding cost parameters are generated by two uniform distributions U[100, 1,000] and U[100, 10,000], representing two different variability levels of the inventory holding cost parameters of all SKU's. The order and ship times from the central warehouse to local warehouses are fixed at 1 day and assumed to be identical for all local warehouses and all SKU's. For the mean repair leadtime at the central warehouse, we use two values ($t_{i,0} = 1$ day for all $i \in I$ and $t_{i,0} = 10$ days for all $i \in I$). Further, two values were used for the target aggregate mean waiting times ($\hat{W}_j^{\text{obj}} = 0.1$ day and $\hat{W}_j^{\text{obj}} = 0.3$ day) and we assume that all local warehouses have the same target. In this test bed, we generate five data samples of the demand rates for each combination of all other parameters (the same inventory holding cost parameters are used for each set of these five data samples). These parameter settings result into 160 instances. Table 6.1 summarizes the parameter settings used in the first test bed.

Test bed 2: The second test bed is the same as the first test bed, apart from the demand rates. We now generate asymmetric demand rates. The same uniform distribution U[0.002, 0.08] is used to generate the demand rate $m_{i,1}$ for each SKU $i \in I$, and we set $m_{i,j} = m_{i,1}$ for all $i \in I$ and $j \in J^{\text{loc}} \setminus \{1\}$. Next, for each $i \in I$ and

Table 6.1 Parameter values for Test bed 1

Parameter	Number of choices	Values
$\|J^{\text{loc}}\|$	2	5, 20
$\|I\|$	2	20, 100
$m_{i,j}$ (per day)	1	U[0.002, 0.08]
c_i^{h} ($ per unit per day)	2	U[100,1,000], U[100, 10,000]
$t_{i,j}$ (in days)	1	1
$t_{i,0}$ (in days)	2	1, 10
$\hat{W}_j^{\text{obj}}$ (in days)	2	0.1, 0.3

$j \in J^{\text{loc}}$, the demand rate $m_{i,j}$ is multiplied by a factor generated from the uniform distribution U[0.2, 2]. Also this test bed consists of 160 instances.

Test bed 3: In the third test bed, we take the instances of the first test bed with five local warehouses ($|J^{\text{loc}}| = 5$), and choose asymmetric targets for the aggregate mean waiting times. In all instances, the target aggregate mean waiting time $\hat{W}_j^{\text{obj}}$ is set at 0.1, 0.15, 0.2, 0.25, 0.3 for $j = 1,2,3,4,5$, respectively. Notice that the number of choices of these targets $\hat{W}_j^{\text{obj}}$ is now reduced to 1. This test bed consists of 40 instances.

Test bed 4: In the fourth test bed, we limit ourselves to the instances with 5 local warehouses, we choose the demand rates as in Test bed 2, and we set the target for the aggregate mean waiting times as in Test bed 3. Hence, again 40 instances are obtained.

For the evaluation of each heuristic $k = 1,2,3,4$, we measure the relative difference between the total costs C_k of the generated heuristic solution and the lower bound C_{MP}:

$$G_k := \frac{C_k - C_{\text{MP}}}{C_{\text{MP}}}, \quad k = 1,2,3,4.$$

6.5.2 Quality of the Heuristics

The results of our experiments for the four test beds are summarized in Tables 6.2–6.5. In each table we present the performance of each heuristic in terms of the average and maximum value of the relative gaps G_k, where we first distinguish subsets of instances with the same value for a specific input parameter and in the bottom line the results for all instances are presented. For example, in Test bed 1 the average and maximum G_k obtained by the greedy approach for all instances with $|J^{\text{loc}}| = 5$ are equal to 7.46 % and 15.35 %, respectively. For the same test bed, the average and maximum G_k obtained by the greedy approach for all 160 instances together are equal to 7.11 % and 23.29 %, respectively. The performance of each heuristic in terms of computation time is presented in Table 6.6. Programs for executing all heuristics were written in MATLAB and all the experiments were run on a PC with a Pentium 4 2.8 GHz processor and 3.37 GB RAM.

Table 6.2 Relative gaps G_k for Test bed 1 (symmetric demand rates, symmetric target aggregate mean waiting times)

Parameter		H1 Greedy		H2 Greedy + LS		H3 DW		H4 DW + LS	
		Avg (%)	Max (%)	Avg (%)	Max (%)	Avg (%)	Max (%)	Avg (%)	Max (%)
$\lvert J^{loc}\rvert$	5	7.46	15.35	5.43	14.62	1.14	7.78	0.61	7.46
	20	6.75	23.29	4.68	14.94	1.00	8.82	0.76	4.68
$\lvert I\rvert$	20	8.28	23.29	5.93	14.94	2.01	8.82	1.29	7.46
	100	5.93	9.65	4.18	8.21	0.13	0.64	0.07	0.35
c_i^h	U[100, 1,000]	6.98	14.37	5.13	14.12	0.90	7.23	0.65	7.23
	U[100, 10,000]	7.24	23.29	4.98	14.94	1.23	8.82	0.72	7.46
$t_{i,0}$	1	7.31	15.35	5.21	14.62	1.30	8.82	0.91	7.46
	10	6.91	23.29	4.90	14.94	0.84	4.51	0.45	3.76
$\hat{W}_j^{obj}$	0.1	6.31	15.04	4.41	11.35	0.92	8.82	0.56	7.46
	0.3	7.91	23.29	5.70	14.94	1.22	7.23	0.80	7.23
All instances		7.11	23.29	5.06	14.94	1.07	8.82	0.69	7.46

Table 6.3 Relative gaps G_k for Test bed 2 (asymmetric demand rates, symmetric target aggregate mean waiting times)

Parameter		H1 Greedy		H2 Greedy + LS		H3 DW		H4 DW + LS	
		Avg (%)	Max (%)	Avg (%)	Max (%)	Avg (%)	Max (%)	Avg (%)	Max (%)
$\lvert J^{loc}\rvert$	5	3.36	12.19	2.86	10.39	1.34	5.03	1.12	4.77
	20	2.55	6.61	2.34	6.35	1.68	5.35	1.50	5.05
$\lvert I\rvert$	20	4.42	12.20	3.86	10.39	2.76	5.35	2.39	5.05
	100	1.49	3.37	1.34	2.89	0.27	0.71	0.23	0.43
c_i^h	U[100, 1,000]	2.79	7.93	2.46	7.91	1.45	5.25	1.33	5.05
	U[100, 10,000]	3.12	12.20	2.74	10.39	1.58	5.35	1.29	4.61
$t_{i,0}$	1	3.70	12.20	3.26	10.39	1.88	5.35	1.62	5.05
	10	2.21	5.39	1.94	5.15	1.15	4.10	1.00	3.42
$\hat{W}_j^{obj}$	0.1	2.60	7.60	2.24	6.57	1.28	4.29	1.07	3.74
	0.3	3.31	12.20	2.96	10.39	1.75	5.35	1.55	5.05
All instances		2.96	12.20	2.60	10.39	1.51	5.35	1.31	5.05

The main observations drawn from these tables can be summarized as follows:

- The DW heuristics (Heuristics 3 and 4) perform very well. In all four test beds the average relative gap G_k is below 2 % and the maximum below 10 %.
- The greedy heuristics (Heuristics 1 and 2) perform also very well in the Test beds 2 and 4 with asymmetric demand rates. The average G_k is around or below 3 % in these test beds, and the maximum is below 12 %. When limiting ourselves to the instances with 100 SKU's in these test beds, we see that the average G_k is even below 2 % and the maximum is below 4 %. The performance of the greedy heuristics in the Test beds 1 and 3 with symmetric demand rates is slightly worse: The average G_k is around or somewhat below 7 %. We think that this phenomenon is due to how the greedy heuristic works. With symmetric demand rates, from our experiment we get the property that if in a given iteration a unit of a specific SKU is stocked at one local warehouse, then also a unit of the same

Table 6.4 Relative gaps G_k for Test bed 3 (symmetric demand rates, asymmetric target aggregate mean waiting times)

Parameter		H1 Greedy		H2 Greedy + LS		H3 DW		H4 DW + LS	
		Avg (%)	Max (%)	Avg (%)	Max (%)	Avg (%)	Max (%)	Avg (%)	Max (%)
$\|I\|$	20	7.81	10.98	6.55	10.04	2.82	5.09	2.36	4.30
	100	4.56	5.97	3.21	4.80	0.32	0.48	0.27	0.43
c_i^{h}	U[100, 1,000]	6.06	9.94	5.09	9.94	1.41	4.78	1.24	4.30
	U[100, 10,000]	6.31	10.98	4.68	10.04	1.73	5.09	1.39	3.97
$t_{i,0}$	1	6.93	10.98	5.78	10.04	1.95	5.09	1.76	4.30
	10	5.44	10.57	3.98	7.71	1.19	3.24	0.87	3.97
All instances		6.19	10.98	4.88	10.04	1.57	5.09	1.32	4.30

Table 6.5 Relative gaps G_k for Test bed 4 (asymmetric demand rates, asymmetric target aggregate mean waiting times)

Parameter		H1 Greedy		H2 Greedy + LS		H3 DW		H4 DW + LS	
		Avg (%)	Max (%)	Avg (%)	Max (%)	Avg (%)	Max (%)	Avg (%)	Max (%)
$\|I\|$	20	4.28	6.78	3.61	6.62	2.68	5.05	2.20	4.87
	100	1.94	2.92	1.61	2.89	0.28	0.40	0.24	0.40
c_i^{h}	U[100, 1,000]	2.99	5.57	2.55	5.02	1.09	3.65	0.99	3.58
	U[100, 10,000]	3.23	6.78	2.66	6.62	1.87	5.05	1.46	4.87
$t_{i,0}$	1	3.79	6.78	3.14	6.62	1.94	5.05	1.60	4.87
	10	2.43	4.20	2.07	3.61	1.01	3.27	0.85	2.44
All instances		3.11	6.78	2.61	6.62	1.48	5.05	1.22	4.87

Table 6.6 Average computation times for each heuristic (seconds)

Parameters		H1 Greedy	H2 Greedy + LS	H3 DW	H4 DW + LS
$\|I\| = 20$	$\|J^{\mathrm{loc}}\| = 5$	4.80	7.24	90.35	92.79
$\|I\| = 20$	$\|J^{\mathrm{loc}}\| = 20$	66.37	119.48	1,132.50	1,185.60
$\|I\| = 100$	$\|J^{\mathrm{loc}}\| = 5$	23.92	38.40	437.05	490.16
$\|I\| = 100$	$\|J^{\mathrm{loc}}\| = 20$	321.50	1,015.12	6,728.24	7,421.95

SKU is stocked at all other local warehouses in the succeeding iterations. This behavior strengthens the discrete character of our optimization problem. Once there is some asymmetry in the demand rates, as one will always have in practical applications, the phenomenon will disappear.

- The improvements obtained by local search are quite limited for both the greedy procedure (compare the results of Heuristics 1 and 2) and the DW method (compare the results of Heuristics 3 and 4).
- The average values of G_k tend to decrease as the problem size (in terms of number of SKU's and local warehouses) or the required stock levels get larger. The latter also occurs when the mean repair leadtime is higher (10 as opposed to 1) or when the target aggregate mean waiting time is lower (0.1 as opposed to 0.3).
- It is shown in Table 6.6 that the greedy method is the most efficient heuristic in terms of computation time. Significant additional computation times are required

when the local search method is applied to improve the solution of the greedy heuristic (compare e.g. 321.50 and 1,015.12 s for the biggest problem size in our experiments). As expected, the computation time of the DW method is considerably higher.

- Computation times increase if the number of SKU's and/or local warehouses increases.

With respect to the quality of the heuristics, we conclude that the greedy procedure (Heuristic 1) is a very appropriate approach for solving the optimization Problem (P) as studied in this chapter. This approach has been proven to be quite effective, and in particular for solving large-size problems, which is the type of problems typically faced in practice. Furthermore, this approach is easy to implement. Adding local search does not lead to much better solutions, while it seriously increases the computational effort. To reduce the computational burden of the procedure, it would be worthwhile to use approximate evaluations instead of the exact method. We analyze this issue in the following section.

6.5.3 Applying Approximate Evaluation Methods

The computational requirements of the exact evaluation method can become rather extensive for instances with a large number of SKU's and/or locations. Using approximate instead of exact evaluations is one way to increase the speed of the greedy procedure. An accurate approximate method will lead us to walk through about the same solutions (as under exact evaluations) while executing the greedy heuristic, and the generated solution will approximately satisfy the aggregate mean waiting time constraints. We know already that Algorithm 6.2, i.e., the approximate method based on two-moment fits is quite accurate; this has been tested in Graves [7] (for the case in which negative binomial distributions are used for the two-moment fits; a similar accuracy will be obtained when using the method of Adan et al. [1] for the two-moment fits). We therefore are particularly interested in evaluating the accuracy of the greedy heuristic when using Algorithm 6.2 for the evaluations. For this purpose, we conduct an experiment using the 320 instances of Test beds 1 and 2. For each instance, we run the greedy heuristic with evaluations by Algorithm 6.2. At the termination of the procedure, the obtained solution is evaluated by the exact method. We record the result with regard to whether or not the generated solution is feasible. If the solution is not feasible, we are also interested in measuring the distance to the set of feasible solutions. For a generated solution **S**, we calculate the relative distance as:

$$\hat{d}(\mathbf{S}) = d(\mathbf{S}) \left(\sum_{j \in J^{\text{loc}}} \hat{W}_j^{\text{obj}} \right)^{-1}$$

Table 6.7 Performance of the greedy heuristic with evaluations by the METRIC approach (Algorithm 6.3)

		Number of instances	Number of feasible solutions	$\hat{d}(\mathbf{S})$		Costs error	
				Avg (%)	Max (%)	Avg (%)	Max (%)
$\lvert J^{\text{loc}}\rvert$	5	160	13	4.56	23.60	2.60	8.62
	20	160	13	1.82	10.03	0.85	2.87
$\lvert I\rvert$	20	160	26	2.71	23.60	1.74	8.62
	100	160	0	3.67	21.11	1.71	7.30
Demand	Symmetric	160	23	3.65	23.60	1.12	6.88
	Asymmetric	160	3	2.74	10.03	2.33	8.62
All instances		320	26	3.19	23.60	1.73	8.62

Table 6.8 Performance of the greedy heuristic with evaluations by Algorithm 6.2

		Number of instances	Number of feasible solutions	$\hat{d}(\mathbf{S})$		Costs error	
				Avg (%)	Max (%)	Avg (%)	Max (%)
$\lvert J^{\text{loc}}\rvert$	5	160	53	0.11	0.75	0.10	1.36
	20	160	62	0.02	0.16	0.02	0.65
$\lvert I\rvert$	20	160	82	0.04	0.49	0.06	1.36
	100	160	33	0.09	0.75	0.06	0.37
Demand	Symmetric	160	71	0.09	0.75	0.04	0.45
	Asymmetric	160	44	0.04	0.16	0.08	1.36
All instances		320	115	0.07	0.75	0.06	1.36

(where $d(\mathbf{S})$ is given by (6.27)). Further, we also measured the costs error, i.e., the relative distance between the heuristic's total costs evaluated by Algorithm 6.2 with the heuristic's total costs obtained by using the exact evaluation method.

Another approximate evaluation method is the METRIC approach (Algorithm 6.3). This method is less accurate, but widely used in practice. In our experiments, we also applied the METRIC approach as an alternative for Algorithm 6.2, so that the effect of using that method is also evaluated.

The results of our computational experiments for the METRIC approach and Algorithm 6.2 are reported in Tables 6.7 and 6.8, respectively. Information on the average computation time under the use of the two approximate evaluation methods and under exact evaluations (as a function of the number of SKU's and local warehouses) is presented in Table 6.9.

The results show us several important points:

- As expected, the METRIC approach is inferior to Algorithm 6.2 with respect to the number of feasible solutions, the distance to the set of feasible solutions, and the costserrors. The METRIC approach gives feasible solutions in only 26 out of 320 data sets. The worst case is observed for data sets with 100 SKU's in which METRIC never gives a feasible solution. The average and maximum relative distances to the set of feasible solutions for METRIC are 3.19 % and 23.6 %. Algorithm 6.2 performs much better by giving 115 feasible solutions with 0.07 % and 0.75 % for the average and maximum relative distances. The average and

Table 6.9 Average computation times for executing the greedy procedure (seconds)

Parameters	METRIC	Algorithm 6.2	Exact
$\lvert I\rvert = 20,\ J^{\text{loc}} = 5$	0.93	2.39	4.80
$\lvert I\rvert = 20,\ J^{\text{loc}} = 20$	6.39	13.33	23.92
$\lvert I\rvert = 100,\ J^{\text{loc}} = 5$	9.49	13.07	66.37
$\lvert I\rvert = 100,\ J^{\text{loc}} = 20$	68.04	84.57	321.50

maximum relative distances between total costs obtained by using the exact and approximate method for METRIC are 1.73 % and 8.62 %. Algorithm 6.2 is quite accurate as indicated by an average error of 0.06 % with a maximum value of 1.36 %.

- We observe too that both methods are more accurate when more local warehouses are involved. For Algorithm 6.2, we can see that all the measures suggest that a higher accuracy is obtained in problems with 20 local warehouses than in problems with 5 local warehouses. For the METRIC approach, we also observe a reduction in relative distances, but the number of feasible solutions remains unchanged. (This is in line with what was pointed out by Axsäter [3]: the METRIC approximation will be more accurate as long as the demand at each local warehouse is low relative to the total demand.)
- The results for both methods, except for the costs errors, deteriorate when dealing with a larger number of SKU's. The distribution of demand across local warehouses also seems to be an influential factor. Both methods are more likely to generate a feasible solution when used for problems with identical demands across local warehouses. In terms of the average relative distance to the set of feasible solutions, both methods are more accurate when used for problems with asymmetric demand rates. The average and maximum total cost error values are higher for problems with asymmetric demand rates.

The results of our experiments show that Algorithm 6.2 is very appropriate to be used within the greedy heuristic. This evaluation method gives highly accurate results and is two to four times faster than exact evaluation. Caution should be taken when using the METRIC approach as it may lead to solutions that are far from the feasible region.

6.6 Concluding Remarks

In this chapter, we studied a multi-item, two-echelon model with backordering of demands when local warehouses are out of stock. This model is almost identical to the well-known METRIC model that was introduced by Sherbrooke [14] in 1968 (see Sect. 6.1 for the precise differences between this chapter's model and the METRIC model).

Sherbrooke [14] himself developed the METRIC approach, an approximate evaluation method based on single-moment fits (see Sect. 6.3.3). Further, he considered the minimization of the total investment in spare parts stock subject to a constraint for the total number of backorders in the system. Later, Simon [16] developed an exact analysis of the system performance under a given basestock policy. This work has been extended to multi-echelon systems by Kruse [9]. Graves [7] developed the exact procedure based on recursive expressions (cf. Sect. 6.3.1), and he developed an approximate evaluation method based on two-moment fits (cf. Sect. 6.3.2). Earlier, Slay [17] developed a similar, but slightly different and more complicated, approximate evaluation method.

Similar to the work of Graves, Rustenburg et al. [13] developed an exact evaluation for multi-echelon, multi-indenture systems; see Chap. 7 for further references on multi-indenture systems. Caggiano et al. [4] developed an exact evaluation for so-called time-based aggregate fill rates in multi-echelon systems, and in addition they developed two heuristic optimization procedures.

Based on either exact or approximate evaluation procedures, one can consider the optimization of inventory holding costs. In this chapter, we looked at the case with an aggregate mean waiting time constraint per local warehouse, and we mainly based ourselves on exact evaluations, like in Wong et al. [20]. We describe a greedy heuristic and a Dantzig-Wolfe heuristic, which are both accurate for problem instances with sufficiently many SKU's. The greedy heuristic is also very efficient; its efficiency is even further increased by using the approximate evaluation method based on the two-moment fits, while this has hardly any effect on the quality of the generated solutions. Hopp et al. [8] and Caglar et al. [5] considered the same problem, but based themselves on approximate evaluations. For an exact optimization method, we refer to Topan et al. [19]. For systems with compound Poisson demand processes, see Stenius et al. [18] and the references therein.

For extensions to two-echelon systems with emergency shipments and lateral transshipments, it is key to have an efficient and accurate approximate evaluation method. When one has only emergency shipments, then the method of Özkan et al. [11] is the best known method. This method outperforms the method of Muckstadt and Thomas [10]. This method of Özkan et al. may also be a good starting point for the analysis of two-echelon systems with emergency shipments and with lateral transshipments within pooling groups.

For more general structures for the emergency and lateral (trans)shipments, there is currently no method available that is efficient and accurate; see also the discussion in Sect. 5.8.

Appendix: Fitting Discrete Distributions on the First Two Moments

In this appendix, we discuss two-moment fits of discrete probability distributions on the first two moments $E\{X\}$ and $E\{X^2\}$ of a random variable X with support $\mathbb{N}_0$.

We first discuss a number of discrete probability distributions. Then we discuss the fit of a negative binomial distribution on the given first two moments, where the so-called stopping parameter k of the negative binomial distribution is allowed to be non-integer. After that we discuss the two-moment fit of Adan et al. [1].

Discrete Probability Distributions

A *Bernoulli trial* is an experiment with two possible outcomes, "success" and "failure". Let p, $0 \le p < 1$, be the probability for the outcome "success". Let $B(p)$ be the random variable that is equal to 1 if the outcome is "success" and 0 if the outcome is "failure". So,

$$P\{B(p)=0\} = 1-p, \;\; P\{B(p)=1\} = p.$$

The distribution of $B(p)$ is called a *Bernoulli distribution* with success probability p. The first two moments and variance are equal to:

$$E\{B(p)\} = p, \;\; E\{(B(p))^2\} = p, \;\; Var\{B(p)\} = p(1-p).$$

Next, suppose that one executes a sequence of k, $k \in \mathbb{N}$, Bernoulli trials, where each trial is independent of earlier trials and each trial has the same probability p for the outcome "success". Let $BIN(k,p)$ be the number of successes. Then

$$P\{BIN(k,p)=x\} = \binom{k}{x} p^x (1-p)^{k-x}, \;\; x \in \mathbb{N}_0.$$

The distribution of $BIN(k,p)$ is called a *binomial distribution* with k trials and success probability p per trial. The first two moments and variance are equal to:

$$E\{BIN(k,p)\} = kp, \;\; E\{(BIN(k,p))^2\} = kp[(k-1)p+1], \tag{6.32}$$
$$Var\{BIN(k,p)\} = kp(1-p). \tag{6.33}$$

Suppose that one executes a sequence of Bernoulli trials, with success probability p per trial, and this sequence is stopped as soon as one gets a failure. Let $GEO(p)$ be the random variable that denotes the number of successes. Then

$$P\{GEO(p)=x\} = p^x(1-p), \;\; x \in \mathbb{N}_0.$$

The distribution of $GEO(p)$ is called a *geometric distribution* with success probability p per trial. The first two moments and variance are equal to:

$$E\{GEO(p)\} = \frac{p}{1-p}, \;\; E\{(GEO(p))^2\} = \frac{p(1+p)}{(1-p)^2},$$
$$Var\{GEO(p)\} = \frac{p}{(1-p)^2}.$$

Let us now consider the execution of a sequence of Bernoulli trials, with success probability p per trial, and assume that this sequence is stopped as soon as one gets the k-th failure. Let $NB(k,p)$ be the random variable that denotes the total number of successes. $NB(k,p)$ is the sum of k independent, geometrically distributed random variables with success probability p, and its distribution is called a *negative binomial distribution* with stopping parameter k and success probability p. The variable $NB(k,p)$ has the above interpretation for strictly positive, integer values of k, but $NB(k,p)$ is also well-defined for strictly positive, non-integer values (i.e., for $k > 0$). The distribution of $NB(k,p)$ is given by

$$P\{NB(k,p) = x\} = \binom{x+k-1}{x} p^x (1-p)^k, \quad x \in \mathbb{N}_0,$$

where the binomial coefficient may be computed by the following formula (for all $k > 0$):

$$\binom{x+k-1}{x} = \frac{(x+k-1)\cdot(x+k-2)\cdot\ldots\cdot k}{x!}.$$

The first two moments and variance are equal to:

$$E\{NB(k,p)\} = \frac{kp}{1-p}, \quad E\{(NB(k,p))^2\} = \frac{kp(1+kp)}{(1-p)^2},$$
$$Var\{NB(k,p)\} = \frac{kp}{(1-p)^2}.$$

Finally, we consider the Poisson distribution. Let $POIS(\lambda)$ be a random variable that is Poisson distributed with parameter $\lambda > 0$. Then:

$$P\{POIS(\lambda) = x\} = \frac{\lambda^x}{x!} e^{-\lambda}, \quad x \in \mathbb{N}_0, \tag{6.34}$$
$$E\{POIS(\lambda)\} = \lambda, \quad E\{(POIS(\lambda))^2\} = \lambda(\lambda+1), \tag{6.35}$$
$$Var\{POIS(\lambda)\} = \lambda. \tag{6.36}$$

Two-Moment Fit by a Negative Binomial Distribution

Let X be a random variable with support $\mathbb{N}_0$ and given first two moments $E\{X\}$ and $E\{X^2\}$. The variance of X is given by $Var\{X\} = E\{X^2\} - (E\{X\})^2$. Assume that

$$Var\{X\} > E\{X\}. \tag{6.37}$$

Notice that this implies that $E\{X\} > 0$ ($E\{X\} = 0$ is only obtained if $X = 0$ with probability 1, but then $Var\{X\}$ would be equal to 0 as well).

Under the assumption of (6.37), we can fit random variable $NB(k,p)$ on the first two moments of X by choosing k and p equal to

$$p = \frac{Var\{X\} - E\{X\}}{Var\{X\}}, \quad k = \frac{1-p}{p} E\{X\}$$

(notice that $0 < p < 1$ and $k > 0$). It is easily seen that then $E\{NB(k,p)\} = E\{X\}$ (this follows immediately from the expression for k) and $Var\{NB(k,p)\} = Var\{X\}$ (use $Var\{NB(k,p)\} = E\{NB(k,p)\}/(1-p) = E\{X\}/(1-p)$ and the expression for p), and thus it also holds that $E\{(NB(k,p))^2\} = E\{X^2\}$.

Two-Moment Fit of Adan et al. [1]

Let X be a random variable with support $\mathbb{N}_0$ and given first two moments $E\{X\}$ and $E\{X^2\}$. We now describe the method of Adan et al. [1] for the fit of a discrete random variable Y (also with support $\mathbb{N}_0$) on these first two moments.

First, determine the squared coefficient of variation $c_X^2 = Var\{X\}/(E\{X\})^2$, and the parameter

$$a = c_X^2 - \frac{1}{E\{X\}}.$$

It holds that $a \geq -1$ (cf. Lemma 2.1 of [1]). We define Y as follows:

(a) If $-1 \leq a < 0$, then choose $k \in \mathbb{N}$ such that $-1/k \leq a < -1/(k+1)$, and

$$Y := \begin{cases} BIN(k,p) & \text{w.p. } q; \\ BIN(k+1,p) & \text{w.p. } 1-q, \end{cases}$$

where

$$q = \frac{1 + a(1+k) + \sqrt{-ak(1+k) - k}}{1+a}, \quad p = \frac{E\{X\}}{k+1-q}.$$

(b) If $a = 0$, then $Y := POIS(\lambda)$ with $\lambda = E\{X\}$.

(c) If $0 < a < 1$, then choose $k \in \mathbb{N}$ such that $1/(k+1) \leq a < 1/k$, and

$$Y := \begin{cases} NB(k,p) & \text{w.p. } q; \\ NB(k+1,p) & \text{w.p. } 1-q, \end{cases}$$

where

$$q = \frac{a(k+1) - \sqrt{(k+1)(1-ak)}}{1+a}, \quad p = \frac{E\{X\}}{k+1-q+E\{X\}}.$$

(d) If $a \geq 1$, then

$$Y := \begin{cases} GEO(p_1) & \text{w.p. } q; \\ GEO(p_2) & \text{w.p. } 1-q, \end{cases}$$

where

$$q = \frac{1}{1+a+\sqrt{a^2-1}}, \quad p_1 = \frac{E\{X\}}{2q+E\{X\}}, \quad p_2 = \frac{E\{X\}}{2(1-q)+E\{X\}}.$$

Then the first two moments of Y are equal to the first two moments of X (cf. Lemma 2.2 of [1]).

The cases (c) and (d) are used when $a > 0$, which is equivalent to (6.37). So, the fits of these cases form an alternative of the two-moment fit by $NB(k, p)$ with a real-valued stopping parameter $k > 0$. The advantage of the fits used in the cases (c) and (d) is that they allow a simple interpretation in terms of mixtures and sums of geometric distributions.

Problems

6.1. Consider Example 6.1. In this example, $W_{i,1}(\mathbf{S}_i)$ has been determined via the exact evaluation procedure, the approximate evaluation procedure based on two-moment fits, and the METRIC approach.

(a) Determine $W_{i,2}(\mathbf{S}_i)$ via the exact evaluation procedure.
(b) Determine $W_{i,2}(\mathbf{S}_i)$ via the approximate evaluation procedure based on two-moment fits. Use a negative binomial distribution for the two-moment fit for $X_{i,2}(S_{i,0})$. How large is the difference with the exact outcome?
(c) Determine $W_{i,2}(\mathbf{S}_i)$ via the METRIC approach. How large is the difference with the exact outcome?

6.2. For a random variable X with support $\mathbb{N}_0$, it is given that $E\{X\} = 1$ and $Var\{X\} = 1.5$.

(a) Fit a negative binomial distribution on the first two moments of X. This gives a negative binomial distribution with specific values for the stopping parameter k and the success probability p. Use the formulas of Appendix "Discrete Probability Distributions" to compute the mean and variance of this negative binomial distribution.
(b) Use the method of Adan et al. to fit a distribution on the first two moments of X (cf. Appendix "Two-Moment Fit of Adan et al. [1]"). Use the formulas of Appendix "Discrete Probability Distributions" to compute the mean and variance of the obtained distribution.

6.3. For the model in this chapter, we have assumed an aggregate mean waiting time constraint per local warehouse. Suppose that the whole set of local warehouses J^{loc} can be partitioned in subsets $J_1^{\text{loc}}, \ldots, J_K^{\text{loc}}$ and that we have an aggregate mean waiting time constraint per subset of local warehouses. This could be relevant when subsets of local warehouses are located in the same region and one wants to control the performance per region rather than per local warehouse. This gives a generalized model.

(a) Formulate the aggregate mean waiting time constraints for this generalized model.

(b) It is easily seen that the evaluation procedures remain applicable without any changes. This does not hold for the heuristics. Formulate an adapted version of the greedy heuristic for the generalized problem.

6.4. In Sect. 6.4.1, it is described that $\Delta_{i,j}(\mathbf{S}) \geq 0$ for all $S \in \mathscr{S}$, $i \in I$ and $j \in J$; see the paragraph after Eqs. (6.29)–(6.30). This can be shown via monotonicity properties for the mean waiting times $W_{i,j}(\mathbf{S}_i)$. In this problem, you are asked to prove these monotonicity properties. Let $i \in I$.

(a) Let $j \in J^{\text{loc}}$, $\mathbf{S} \in \mathscr{S}$, and $\mathbf{S}' = \mathbf{S} + \mathbf{E}_{i,j}$. Show that

- $W_{i,j}(\mathbf{S}') \leq W_{i,j}(\mathbf{S})$;
- $W_{i,j'}(\mathbf{S}') = W_{i,j'}(\mathbf{S})$ for all $j' \in J^{\text{loc}}$, $j' \neq j$.

(b) Let $\mathbf{S} \in \mathscr{S}$, and $\mathbf{S}' = \mathbf{S} + \mathbf{E}_{i,0}$. Show that $W_{i,j}(\mathbf{S}') \leq W_{i,j}(\mathbf{S})$ for all $j \in J^{\text{loc}}$.

Hint: Use the recursive expressions of Sect. 6.3.1.

6.5. Consider a given SKU $i \in I$ and assume that at the central warehouse an (s,Q)-policy instead of a basestock policy is used for this SKU. Its reorder level is given by s_i and the fixed batch size by Q_i. We study how the exact evaluation method can be adapted for this generalized inventory policy.

(a) Derive adapted formulas for the distribution of the on-hand stock $OH_{i,0}$ and number of backorders $BO_{i,0}$ at the central warehouse. Hint: Use the results of Sect. 2.10.3.
(b) Denote what needs to be adapted in the rest of the exact evaluation procedure.

6.6. Consider an OEM with a two-echelon spare parts system as described in this chapter. The OEM has service contracts with many customers regarding the spare parts supply. Each customer is coupled to a local warehouse and the OEM ensures that the aggregate mean waiting time per local warehouse stays within a certain limit. Customers without service contracts can also buy spare parts, but they are served from the central warehouse. Their orders constitute a so-called *direct demand stream* at the central warehouse. For each SKU $i \in I$, this extra demand stream is assumed to be a Poisson stream and its demand rate is denoted by $\hat{m}_i$. At the central warehouse, one treats these demands in the same way as replenishment orders from the local warehouses.

The above situation leads to a generalized model with a direct demand stream at the central warehouse. The formulation of Problem (P) remains the same under this generalization. Describe what needs to be changed in the exact evaluation procedure to deal with this generalization.

6.7. Six European countries plan to buy the same military system. We consider a single, expensive repairable for which all repairs are executed by the original manufacturer from the US. We consider the following two scenarios:

- *Scenario 1:* Each country organizes its own spare parts inventory. In that case, each country has a own stockpoint and failed parts are sent back to the manufacturer in the US. After repair, the same parts come back as ready-for-use parts. The leadtime for this repair loop is 6 months.

- *Scenario 2:* The countries create pooling via a joint central warehouse. Countries send their failed parts back to the central warehouse (within a negligibly small leadtime), and request a ready-for-use part from there. The central warehouse sends the failed parts back to the manufacturer in the US, and after repair they are returned to the central warehouse as ready-for-use parts. This repair loop has a leadtime of 6 months (as in scenario 1). The leadtime for sending parts from the central warehouse to a local warehouse in a country is 1 week (= 0.02 years, say).

For Scenario 2, further assumptions are such that the model of Chap. 6 applies (with 1 SKU instead of a general number of SKU's). For Scenario 1, similar assumptions are made. Per country, the total failure rate for the repairable is 2 failures per year. Each country has a target of 0.01 years for the mean waiting time per failure.

The minimum amount of spare parts that each country has to take on stock under Scenario 1 is denoted by $\hat{S}$. The total amount of stock under Scenario $i = 1,2$ is denoted by $S_{\text{tot}}^{(i)}$.

Under Scenario 2, one has more transportation movements and more handling because of the central warehouse. However, the extra costs are limited because the central warehouse also leads to consolidation of the transport to and from the US. Further, the repairable is expensive and thus inventory holding costs dominate. Hence, it is justified to compare both scenarios on the basis of the total stock that is required to meet the mean waiting time constraints.

(a) For Scenario 1, determine $\hat{S}$ and $S_{\text{tot}}^{(1)}$. What is the mean waiting time under these values?
(b) Next, we analyze Scenario 2, and assume that a basestock level of 1 is used in each local warehouse. Initially, we consider this scenario with a total stock $S_{\text{tot}}^{(2)}$ that is equal to the total stock $S_{\text{tot}}^{(1)}$ found under (a), i.e., a basestock level equal to $S_{\text{tot}}^{(1)} - 6$ is used at the central warehouse. Determine the mean waiting time that is obtained at each of the local warehouses under this basestock policy?
(c) Suppose that one sticks to a basestock level of 1 at each of the local warehouses. What is the minimum basestock level at the central warehouse that one needs to meet the mean waiting time constraints? What is the total stock under this value, and how large is the reduction in total stock compared to the total stock required under scenario 1?

References

1. Adan, I.J.B.F., Van Eenige, M.J.A., Resing, J.A.C.: Fitting discrete distributions on the first two moments. Probab. Eng. Inf. Sci. **9**, 623–632 (1996)
2. Axsäter, S.: Simple solution procedures for a class of two-echelon inventory systems. Oper. Res. **38**, 64–69 (1990)
3. Axsäter, S.: Continuous review policies for multi-level inventory systems with stochastic demand. In: Graves, S.C., Rinnooy Kan, A.H.G., Zipkin, P.H. (eds.) Logistics of Production and Inventory, Chap. 10. North-Holland, New York (1993)

4. Caggiano, K.E., Jackson, P.L., Muckstadt, J.A., Rappold, J.A.: Optimizing service parst inventory in a multiechelon, multi-item supply chain with time-based customer service-level agreements. Oper. Res. **55**, 303–318 (2007)
5. Caglar, D., Li, C.L., Simchi-Levi, D.: Two-echelon spare parts inventory system subject to a service constraint. IIE Trans. **36**, 655–666 (2004)
6. Dantzig, G.B., Wolfe, P.: Decomposition principle for linear programs. Oper. Res. **8**, 101–111 (1960)
7. Graves, S.C.: A multi-echelon inventory model for a repairable item with one-for-one replenishment. Manag. Sci. **31**, 1247–1256 (1985)
8. Hopp, W.J., Zhang, R.Q., Spearman, M.L.: An easily implementable hierarchical heuristic for a two-echelon spare parts distribution system. IIE Trans. **31**, 977–988 (1999)
9. Kruse, K.C.: An exact *N* echelon inventory model: the simple Simon method. U.S. Army Research Office, Technical Report TR 79-2 (1979)
10. Muckstadt, J.A., Thomas, L.J.: Are multi-echelon inventory methods worth implementing in systems with low-demand-rate items? Manag. Sci. **26**, 483–494 (1980)
11. Özkan, E., Van Houtum, G.J., Serin, Y.: A new approximate evaluation method for two-echelon inventory systems with emergency shipments. Ann. Oper. Res. **224**, 147–169 (2015)
12. Palm, C.: Analysis of the Erlang traffic formula for busy-signal arrangements. Ericsson Tech. **4**, 204–212 (1938)
13. Rustenburg, W.D., Van Houtum, G.J., Zijm, W.H.M.: Exact and approximate analysis of multi-echelon, multi-indenture spare parts systems with commonality. In: Shanthikumar, J.G., Yao, D.D., Zijm, W.H.M. (eds.) Stochastic Modeling and Optimization of Manufacturing Systems and Supply Chains, pp. 143–176. Kluwer Academic, Boston/Dordrecht/London (2003)
14. Sherbrooke, C.C.: METRIC: A multi-echelon technique for recoverable item control. Oper. Res. **16**, 122–141 (1968)
15. Sherbrooke, C.C.: Optimal inventory modeling of systems: multi-echelon techniques. Kluwer Academic, Boston/Dordrecht/London (2004)
16. Simon, R.M.: Stationary properties of a two-echelon inventory model for low demand items. Oper. Res. **19**, 761–773 (1971)
17. Slay, F.M.: VARI-METRIC: an approach to modeling multi-echelon resupply when the demand process is Poisson with a Gamma prior. Report AF301-3, Logistics Management Institute, Washington, DC (1984)
18. Stenius, O., Karaarslan, A.G., Marklund, J., De Kok, A.G.: Exact analysis of divergent inventory systems with time-based shipment consolidation and compound Poisson demand. Working paper, Lund University (2014)
19. Topan, E., Bayindir, Z.P., Tan, T.: An exact solution procedure for multi-item two-echelon spare parts inventory control problem with batch ordering in the central warehouse. Oper. Res. Lett. **38**, 454–461 (2010)
20. Wong, H., Kranenburg, A.A., Van Houtum, G.J., Cattrysse, D.: Efficient heuristics for two-echelon spare parts inventory systems with an aggregate mean waiting time constraint per local warehouse. OR Spectr. **29**, 699–722 (2007)[1]

[1] The material of this paper has been reused in this chapter with kind permission from Springer Science+Business Media.

Chapter 7
Multi-echelon, Multi-indenture System

7.1 Introduction

In Chap. 6, we assumed that all failed components were sent back to the central warehouse and that they were repaired at internal and external repair shops at the central warehouse. In this chapter, we generalize the model of Chap. 6 as follows:

- We assume a general distribution network with a central point, locations at close distance of installed machines, and intermediate points.
- We assume that the machines have a so-called *multi-indenture* structure (cf. Sherbrooke [14], who introduced the concept of indenture levels). This means that machines consist of critical components at multiple levels of their material breakdown. The components that are replaced when a failure occurs, are called *assemblies*. For the repair of assemblies, one may have to replace an underlying component, which is called a *subassembly*; and for the repair of a subassembly one may need a next underlying component, and so on. Hence, it is relevant to keep spare parts on stock at the level of assemblies, the level of subassemblies, and lower levels.
- We assume that repair of assemblies and lower level components is possible at multiple locations in the distribution network. Because of this, each location has a storage function for spare parts and it may have repair facilities for the repair of failed parts. We therefore refer to the points in the distribution network as locations rather than warehouses.

The model that we get by these generalizations is a multi-echelon, multi-indenture model. For the material breakdown of the machines, we allow that the subassemblies occur in multiple assemblies, and similarly for lower level components. This is a form of *commonality*. Obviously, as before we also allow commonality at the level of the assemblies. We assume that the installed machines at the various locations are of the same type, and they have commonality at the level of the assemblies.

G.J. van Houtum, B. Kranenburg, *Spare Parts Inventory Control under System Availability Constraints*, International Series in Operations Research & Management Science 227, DOI 10.1007/978-1-4899-7609-3_7

The multi-echelon, multi-indenture models are relevant for situations in practice where repairs of assemblies get significantly delayed when required subassemblies or lower level parts are not available in ample amounts. It is then important to balance the stocks of assemblies, subassemblies, and lower level components. They are also appropriate for a joint consideration of decisions on the spare parts stock, on where components of different levels should be repaired, and on where possibly investments in repair equipment are needed; see [2, 4, 5].

For the multi-echelon, multi-indenture model of this chapter, we generalize the exact evaluation and the approximate evaluation based on two-moments as described in Chap. 6 for the two-echelon model. The approximate evaluation based on two-moment fits is accurate and efficient, and is a good method to be used in heuristic optimization methods. We will describe a greedy heuristic, which worked well for the two-echelon model in Chap. 6. It is reasonable to assume that a greedy heuristic works also well for the generalized model of this chapter. We also present a case study at the Royal Netherlands Navy, executed in 2000 (see Rustenburg [11]).

The organization of this chapter is as follows. In Sect. 7.2, we describe our model. Next, in the Sects. 7.3 and 7.4, we describe the exact and approximate evaluation. Then, in Sect. 7.5, we described the greedy heuristic. Finally, we present the case study at the Royal Netherlands Navy in Sect. 7.6, and we make concluding remarks in Sect. 7.7.

This chapter is largely based on Rustenburg et al. [12].

7.2 Model

We start the model description with discussing the presence of assemblies and lower level components and the divergent multi-echelon network of locations where spare parts are stocked and repaired. After that, we describe the repair and distribution process in more detail and give an overview of assumptions and notations.

We assume the existence of a set J^{ba} of locations, also called *bases*, each of which serves a number of machines. Machines at the same base are assumed to be of the same machine type. A machine consists of multiple critical components, which are subject to failures. These components occur at multiple levels in the material breakdown structure. Among these components, we have so-called *assemblies*, which are components that can be replaced as a whole. When a machine fails, then this is first observed at the level of an assembly. Later on, when the assembly is repaired, one can identify whether one of the subassemblies was the cause of the failure. Similarly, when a subassembly is repaired, one can identify whether one of the components at the next level was the cause of the failure, and so on.

So, initially, demands for spare parts occur at the level of assemblies. Let I^{as} denote the set of all assemblies that may occur in the configurations of the machines. We assume that the total demand stream for parts of an assembly i at a base j constitutes a Poisson process with a constant rate $m_{i,j}$. For many real-life systems, lifetimes of assemblies are (close-to-)exponential, or lifetimes are not precisely exponential but the total stream of failures is a composition of subprocesses coming from relatively many machines that are supported by a base. In those cases it is reasonable to assume a Poisson demand process. Further, in practice, one does not allow long downtimes of machines, and thus, for relevant situations, it is reasonable to assume constant demand rates. For each base $j \in J^{\text{ba}}$, we define $M_j := \sum_{i \in I^{\text{as}}} m_{i,j}$ as the total demand rate for assemblies at base j; we assume that $M_j > 0$.

Let I denote the set of all assemblies, subassemblies, and lower level components. Hence, $I \supset I^{\text{as}}$. We denote I also as the set of all Stock-Keeping Units (SKU's). The SKU's are numbered $1, \ldots, |I|$ ($|I| \in \mathbb{N}$). Each SKU $i \in I \setminus I^{\text{as}}$ has a set of parents $P_{\text{mb}}(i)$. We allow *commonality*, i.e., subassemblies may occur in the configurations of two or more assemblies, and similarly for lower level SKU's. So, we may have $|P_{\text{mb}}(i)| > 1$ for several SKU's $i \in I \setminus I^{\text{as}}$. For each SKU $i \in I$, $C_{\text{mb}}(i) = \{k \in I | i \in P_{\text{mb}}(k)\}$ denotes the set of children of i. The set $I^{\text{cl}} = \{i \in I | C_{\text{mb}}(i) = \emptyset\}$ consists of all childless SKU's. We assume that the directed graph with the SKU's as nodes and the parent-child relations as directed arcs contains no cycles. W.l.o.g., we assume that the assemblies and underlying SKU's are numbered such that $k < i$ for all $k \in P_{\text{mb}}(i)$, $i \in I \setminus I^{\text{as}}$.

Notice that very general material breakdown structures can be modeled. Also single-indenture systems fit. For those systems, we have $I^{\text{as}} = I^{\text{cl}} = I$.

Each base location may hold stock of any SKU and in addition may have repair facilities. Apart from the bases, there exists a number of support locations with the same possibilities. These locations in turn may be supported by a few other locations, and so on. The network of all locations is supposed to constitute a tree. Hence, there is a single root location in the network. This station is supported by external suppliers. Each of the other locations has a *unique* supporting location. Let J denote the set of all locations. Hence, $J \supset J^{\text{ba}}$. The root location, also called *central depot*, is denoted as location 0 and the other locations are numbered $1, \ldots, |J| - 1$. For each location $j \geq 1$, the unique supporting location, called the parent, is denoted by $p_{\text{dn}}(j)$. For computational convenience, we assume that the locations are numbered such that $p_{\text{dn}}(j) < j$, $j \in J \setminus \{0\}$. For each location j, $C_{\text{dn}}(j) = \{l \in J | p_{\text{dn}}(l) = j\}$ denotes the set of children. Hence, the set of base locations J^{ba} satisfies $J^{\text{ba}} = \{j \in J | C_{\text{dn}}(j) = \emptyset\}$. Notice that the distribution network may consist of any number of echelon levels. Also the extreme case with only one echelon level is allowed. In that case, we have one single location only, which is a base and central depot at the same time. Then $J = J^{\text{ba}} = \{0\}$.

Example 7.1. We consider the example of a fire extinguishing system, of which identical units are installed at 5 bases. There is one unit per base. The bases are supported by a central depot. The material breakdown and distribution network are depicted in Figs. 7.1 and 7.2, respectively.

Each fire extinguishing system consists of 2 different pump units, which are the assemblies. They are numbered 1 and 2. Both pump units consist of a pump and an electromotor. The electromotors are of different types, but the two pumps in the units are identical. Hence, the pumps and all their components are common. In Fig. 7.1, one can see lower level components that occur in the pump and two different electromotors. In total, we have 12 SKU's, with numbers according to

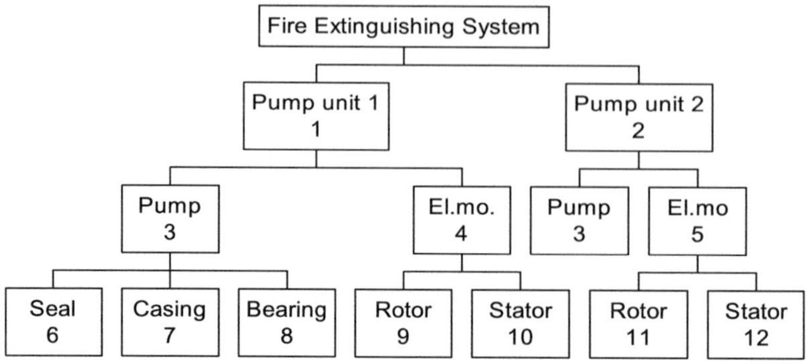

Fig. 7.1 The material breakdown structure of the fire extinguishing system of Example 7.1

Fig. 7.1. It holds that $|I| = 12$, $I^{\text{as}} = \{1,2\}$, $I^{\text{cl}} = \{6,\ldots,12\}$; for the sets of parents $P_{\text{mb}}(i)$ and sets of children $C_{\text{mb}}(i)$, we refer to Fig. 7.1.

For the distribution network, it holds that the bases are numbered as $1,\ldots,5$. It holds that $|J| = 6$, $J^{\text{ba}} = \{1,\ldots,5\}$, $p_{\text{dn}}(1) = \ldots = p_{\text{dn}}(5) = 0$, $C_{\text{dn}}(0) = \{1,\ldots,5\}$, and $C_{\text{dn}}(1) = \ldots = C_{\text{dn}}(5) = \emptyset$.

We now describe the operational process, which involves the failures, repairs and distribution of parts. A series of actions starts with each failure of a machine. Because failures are assumed to be observed at the assembly level only, the description starts with a failure of an assembly.

Suppose a part of assembly $i \in I^{\text{as}}$ of a machine at some base location $j \in J^{\text{ba}}$ fails. Then the machine goes down. This part is replaced by a spare part of the stock at location j if available, while the failed part is sent into repair. With probability $r_{i,j}$, the part can be repaired at base j, and then the part is sent to the repair shop at the base. With probability $1 - r_{i,j}$, the part cannot be repaired at base j. In the latter case, the failed part is sent to the parent location $p_{\text{dn}}(j)$, while at the same time a replenishment order for one unit of assembly i is placed at location $p_{\text{dn}}(j)$. The order and ship time for a part i from the parent location $p_{\text{dn}}(j)$ to j is equal to $T_{i,j}^{\text{res}}$ (a random variable, where the superscript "res" stands for resupply; later we specify which order and ship times are assumed to be deterministic and which order and ship times are allowed to be stochastic). This time is excluding a possible delay at location $p_{\text{dn}}(j)$ in case the requested spare part is not immediately available there. In case the failed part can be repaired at base j, the repair action involves the possible detection of a subassembly k that causes the problem. Let $q_{i,k,j}$ denote the probability that the failure of assembly i is due to subassembly k. With probability $1 - \sum_{k \in C_{\text{mb}}(i)} q_{i,k,j}$ the failure is not due to one of the children $k \in C_{\text{mb}}(i)$. Then the failure may be due to a component that is not considered in our model and of which always sufficient spare parts are available, or the failure is due to environmental conditions (for instance, dust), or a real repair of the part itself is needed. Once part i is sent into repair at base j (i.e., *after* a possible delay because a requested subassembly is not available immediately), it takes a repair leadtime $T_{i,j}^{\text{rep}}$

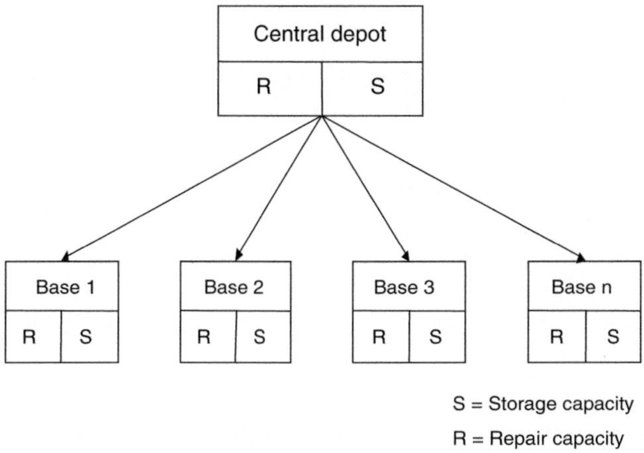

Fig. 7.2 The distribution network of Example 7.1

until the part is returned to the spare parts stock as a ready-for-use part. If the repair of the part is done at the parent location $p_{\text{dn}}(j)$, then the same actions take place at location $p_{\text{dn}}(j)$.

A failed part of subassembly k undergoes the same routine, i.e. after inspection a possible failed lower level part is detected and replaced. Again, repair of the subassembly may not be possible at the location itself in which case the parent location becomes involved. In this way, we may proceed along the material breakdown structure of any assembly until its lowest level is reached. In principle at each level, except the lowest one, a repair action basically consists of a disassembly, a part replacement and finally an assembly action. Only at the lowest level repair indeed means an actual repair action. At any level it may be that repair is not possible at the base in which case the next higher location is asked for support, i.e., the failed part is sent there and a ready-for-use part is shipped downstream to replenish the stock. The higher location in turn may need help as well from its supporting location, and so on. However, if a part of a SKU i appears to be irreparable at the central depot, which happens with probability $1 - r_{i,0}$, then it is disposed of. Then the part is called *condemned*, and immediately a new part is ordered at an external supplier in order to replace this condemned part. We assume that the external supplier of SKU i can always deliver. The total procurement time is denoted by the random variable $T_{i,0}^{\text{res}}$. In the rest of this chapter, we see and treat the procurement times as order and ship times, and we also denote them as order an ship times.

Notice that implicitly a *one-for-one replenishment and repair policy* has been assumed for all SKU's at all locations. I.e., each SKU i at each location j is controlled according to a *basestock policy*. The corresponding basestock level is denoted by $S_{i,j}$. Since we often have to deal with expensive slow moving SKU's this seems to be a reasonable assumption. Let $\mathbf{S}_i := (S_{i,0}, S_{i,1}, \ldots, S_{i,|J|-1})$, $i \in I$, denote the vector of basestock levels for SKU i, and let a basestock policy be denoted by

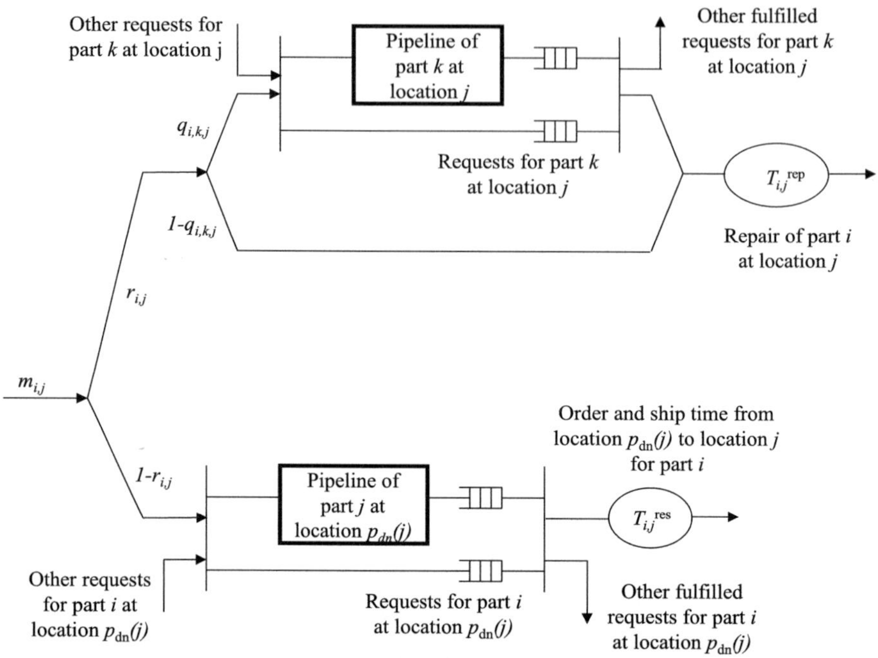

Fig. 7.3 Possible actions for a failed part $i \in I$ with one child $k \in C_{\text{mb}}(i)$ at a location $j \in J \setminus \{0\}$

$$\mathbf{S} = \begin{pmatrix} S_{1,0} & S_{1,1} & \dots & S_{1,|J|-1} \\ S_{2,0} & S_{2,1} & \dots & S_{2,|J|-1} \\ \vdots & \vdots & \ddots & \vdots \\ S_{|I|,0} & S_{|I|,1} & \dots & S_{|I|,|J|-1} \end{pmatrix}.$$

Also note that consumable SKU's are included in our model. Consumables SKU's are irreparable at any location (and thus they are disposed of at the central depot), i.e., for a consumable SKU i we simply have $r_{i,j} = 0$ for all $j \in J$.

Obviously, for each part $i \in I$ at each location $j \in J$ backordered demands may occur. In all cases these backorders are treated in FCFS order. Possible actions for a failed part $i \in I$ with one child $k \in C_{\text{mb}}(i)$ at a location $j \in J \setminus \{0\}$ are visualized in Fig. 7.3. Similar pictures are obtained for other SKU's and for possible actions at the central depot.

With respect to the repair leadtimes $T_{i,j}^{\text{rep}}$ and the order and ship times $T_{i,j}^{\text{res}}$, we assume the following. We assume that the expected values of all these times are given. For each $i \in I$, $j \in J$, the mean of $T_{i,j}^{\text{rep}}$ and $T_{i,j}^{\text{res}}$ is denoted by $t_{i,j}^{\text{rep}}$ and $t_{i,j}^{\text{res}}$, respectively. For all SKU's i, the order and ship times $T_{i,0}^{\text{res}}$ at the central depot may be stochastic. The order and ship times of subsequent replenishment orders for a SKU $i \in I$ placed by the central depot are assumed to be independent and identically distributed, and we assume that times for different SKU's are mutually independent. Further, for all childless parts $i \in I^{\text{cl}}$, the repair leadtimes $T_{i,j}^{\text{rep}}$ may be stochastic at all location $j \in J$. Here we also assume independent and identically distributed times and mutually independency for different SKU's. All other repair leadtimes and all other order and ship times are assumed to be *deterministic*. Under these assumptions, an exact evaluation is possible as we will see in Sect. 7.3. For the approximate evaluation, as presented in Sect. 7.4, we may allow stochastic order and ship times and stochastic repair leadtimes for all SKU's at all locations; see Remark 7.1.

The objective is to minimize total inventory holding costs subject to aggregate mean waiting time constraints for assemblies at the base locations. For the inventory holding costs, we count all parts on stock in the network, all parts in transport, and all parts on stock. For SKU i, the inventory holding costs are c_i^{h} (> 0) per part per unit of time. Let $C_i(\mathbf{S}_i)$ be the expected total costs per time unit for SKU i, $i \in I$:

$$C_i(\mathbf{S}_i) = c_i^{\mathrm{h}} \sum_{j \in J} S_{i,j}. \tag{7.1}$$

The total average costs are equal to $C(\mathbf{S}) = \sum_{i \in I} C_i(\mathbf{S}_i)$. At each base $j \in J^{\mathrm{ba}}$, there is a maximum level $\hat{W}_j^{\mathrm{obj}}$ given for the aggregate mean waiting time for requested assemblies. Under a basestock policy $\mathbf{S}$, the aggregate mean waiting time $\hat{W}_j(\mathbf{S})$ for assemblies at base j is equal to

$$\hat{W}_j(\mathbf{S}) = \sum_{i \in I^{\mathrm{as}}} \frac{m_{i,j}}{M_j} W_{i,j}(\mathbf{S}), \tag{7.2}$$

where $W_{i,j}(\mathbf{S})$ denotes the mean waiting time for a demand for assembly i at base j (notice that this mean waiting time does not only depend on the basestock levels for i itself, but also on basestock levels of other SKU's). Our optimization problem is formulated as:

$$\begin{aligned} (\mathrm{P}) \qquad \min \qquad & C(\mathbf{S}) \\ \text{subject to } & \hat{W}_j(\mathbf{S}) \le \hat{W}_j^{\mathrm{obj}}, \quad j \in J^{\mathrm{ba}} \\ & \mathbf{S} \in \mathscr{S}, \end{aligned}$$

where $\mathscr{S} = \{\mathbf{S} \mid S_{i,j} \in \mathbb{N}_0, \forall i \in I \text{ and } j \in J\}$. An optimal policy of Problem (P) is denoted by $\mathbf{S}^*$ and the optimal costs are denoted by C_{P}.

7.2.1 Overview of Assumptions and Notations

Because the model is quite advanced, we list both the main assumptions and the notation. The main assumptions are as follows:

(a) At each of the bases, the failures for the different assemblies occur according to independent Poisson processes;
(b) Each failure of a part with one or more children is due to at most one child;

(c) The directed graph with the components as nodes and the parent-child relations as directed arcs contains no cycles. Further, no assembly occurs in the material breakdown structure of another assembly;
(d) For each childless SKU, at all locations the repair leadtimes for that SKU are independent and identically distributed random variables. For all other SKU's, the repair leadtimes are assumed to be deterministic;
(e) For each SKU, the order and ship times at the central depot are independent and identically distributed random variables. For all other locations, all order and ship times are assumed to be deterministic;
(f) A one-for-one replenishment and repair policy is applied for all SKU's at all locations;
(g) There are no lateral transshipments or emergency shipments in the distribution network.

We have the following notation for our model:

- I: Set of SKU's, SKU's are numbered $1,\ldots,|I|$;
- I^{as}: Set of assemblies;
- I^{cl}: Set of childless SKU's;
- $P_{\text{mb}}(i)$, $i \in I \setminus I^{\text{as}}$: Set of parents of SKU i. The SKU's are numbered such that $k < i$ for all $k \in P_{\text{mb}}(i)$;
- $C_{\text{mb}}(i)$, $i \in I$: Set of children of SKU i;
- J^{ba}: Set of bases, the locations where machines are installed;
- J: Set of all locations, locations are numbered $0,1,\ldots,|J|-1$, index 0 is used for the central depot;
- $p_{\text{dn}}(j)$, $j \in J \setminus \{0\}$: Unique parent location of location j. The locations are numbered such that $p_{\text{dn}}(j) < j$ for all $j \in J \setminus \{0\}$;
- $C_{\text{dn}}(j)$, $j \in J$: Set of children of location j;
- $m_{i,j}$: Demand rate for SKU i at location j (input variable for $i \in I^{\text{as}}$ and $j \in J^{\text{ba}}$; for other combinations of i and j, this variable is computed from the input variables; see Sect. 7.3);
- $M_j = \sum_{i \in I^{\text{as}}} m_{i,j}$, $j \in J^{\text{ba}}$: Total demand rate for assemblies at base j;
- $r_{i,j}$: Probability that a failed part of SKU i is repairable at location $j \in J$;
- $q_{i,k,j}$, $i \in I, k \in C_{\text{mb}}(i)$, $j \in J$: Probability that a failed part of SKU i that is repaired at location j, needs a ready-for-use part of SKU $k \in C_{\text{mb}}(i)$ for the repair;
- $T^{\text{rep}}_{i,j}$: Repair leadtime for a part of SKU i that is repaired at location j;
- $t^{\text{rep}}_{i,j}$: Mean of repair leadtime $T^{\text{rep}}_{i,j}$;
- $T^{\text{res}}_{i,j}$: Order and ship time for a part of SKU i that is being sent to location j from location $p_{\text{dn}}(j)$ if $j \in J \setminus \{0\}$ and from an external supplier if $j = 0$;
- $t^{\text{res}}_{i,j}$: Mean of the order and ship time $T^{\text{res}}_{i,j}$;
- $S_{i,j}$: Basestock level for SKU i at location j;
- $\mathbf{S}_i = (S_{i,0}, S_{i,1},\ldots,S_{i,|J|-1})$: Basestock vector for SKU i;
- $\mathbf{S}$: $|I| \times |J|$ matrix with all basestock levels $S_{i,j}$, denotes a basestock policy;
- $\mathscr{S}$: Solution space, set of all possible basestock policies $\mathbf{S}$;
- c^{h}_i: Inventory holding cost per time unit per unit of SKU i;
- $C_i(\mathbf{S}_i)$: Average costs for SKU i;

Table 7.1 Input data for Example 7.1

SKU no.	Name	$m_{i,j}$ (p.year)	$r_{i,j}$	$r_{i,0}$	$t^{\text{rep}}_{i,j}$ (years)	$t^{\text{rep}}_{i,0}$ (years)	$t^{\text{res}}_{i,j}$ (years)	$t^{\text{res}}_{i,0}$ (years)	c^{h}_i (Euro/year)
1	p.unit 1	20.4	0.8	0.95	0.01	0.1	0.2	0.75	550
3	Pump	–	0.2	0.7	0.03	0.2	0.2	0.5	99
6	Bearing	–	0.2	0	0.1	–	0.2	0.3	16.5
7	Seal	–	0.2	0	0.1	–	0.2	0.3	22.5
8	Casing	–	0.2	0	0.1	–	0.2	0.3	22
4	Elmo	–	0.2	0.75	0.03	0.2	0.2	0.5	254
9	Rotor	–	0.2	0	0.1	–	0.2	0.3	7.5
10	Stator	–	0.2	0	0.1	–	0.2	0.3	24
2	p.unit 2	13.6	0.8	0.95	0.01	0.1	0.2	0.75	500
5	Elmo	–	0.2	0.75	0.03	0.2	0.2	0.5	165
11	Rotor	–	0.2	0	0.1	–	0.2	0.3	22.5
12	Stator	–	0.2	0	0.1	–	0.2	0.3	22

Table 7.2 The failure probabilities $q_{i,k,j}$ for Example 7.1 (the $q_{i,k,j}$ are the same for all $j \in J$)

$i \backslash j$	3	4	5	6	7	8	9	10	11	12
1	0.55	0.45								
2	0.38		0.62							
3				0.32	0.47	0.21				
4							0.29	0.71		
5									0.37	0.63

- $C(\mathbf{S}) = \sum_{i \in I} C_i(\mathbf{S}_i)$: Total average costs;
- $W_{i,j}(\mathbf{S})$, $i \in I^{\text{as}}$, $j \in J^{\text{ba}}$: Mean waiting time for assembly $i \in I^{\text{as}}$ at base $j \in J^{\text{ba}}$;
- $\hat{W}_j(\mathbf{S})$, $j \in J^{\text{ba}}$: Aggregate mean waiting time for an assembly at base $j \in J^{\text{ba}}$;
- $\hat{W}^{\text{obj}}_j$, $j \in J^{\text{ba}}$: Target level for the aggregate mean waiting time at base $j \in J^{\text{ba}}$;
- $\mathbf{S}^*$: An optimal solution of Problem (P);
- C_{P}: Optimal costs of Problem (P).

Example 7.1 (continued). We assume symmetry in the input data for the various bases, i.e., identical values are assumed for the $m_{i,j}$, $r_{i,j}$, $t^{\text{rep}}_{i,j}$, $t^{\text{res}}_{i,j}$, and $q_{i,k,j}$ for all bases $j \in J^{\text{ba}}$. The input data are given in Tables 7.1 and 7.2. The $q_{i,k,j}$ are given in the latter table. The $q_{i,k,0}$ at the central depot are assumed to be the same as the $q_{i,k,j}$ at the bases. Notice that the repair probability $r_{i,0}$ at the central depot equals 0 for the SKU's $6, 7, \ldots, 12$; for that reason no mean repair leadtimes $t^{\text{rep}}_{i,0}$ are given for these SKU's.

7.3 Exact Evaluation

In this section we present an exact performance analysis. First we give some preliminary results. After that we derive recursive expressions for the numbers of backorders and numbers of parts in repair and on order. Based on these recursive procedures, the evaluation procedure is easily deduced.

7.3.1 Preliminary Results

In the initial state of the spare parts network we have an initial number of $S_{i,j}$ spare parts on stock of each SKU $i \in I$ at each location $j \in J$. These SKU's are demanded according to certain demand processes and the stock is replenished according to a basestock policy with the $S_{i,j}$ as basestock levels. The latter means that the inventory position of each SKU $i \in I$ at each location $j \in J$ is kept at a constant level $S_{i,j}$.

First, we look at the demand processes. The demands for an assembly $i \in I^{\rm as}$ at a base $j \in J^{\rm ba}$ occur according to a Poisson process with a given rate $m_{i,j}$. Each demand may immediately result in a demand for a subassembly at base j or a demand for a unit of assembly i at the parent location $p_{\rm dn}(j)$. A demand for assembly i at base j immediately results in a demand for a subassembly $k \in C_{\rm mb}(i)$ with probability $r_{i,j}q_{i,k,j}$ and in a demand for a unit of assembly i at location $p_{\rm dn}(j)$ with probability $1 - r_{i,j}$. As a result, the Poisson demand process for assembly i at base j splits into independent Poisson demand processes for subassemblies at the same base and the same assembly at the parent location. These subprocesses join with independent other subprocesses and thus constitute Poisson demand processes for the subassemblies $k \in C_{\rm mb}(i)$ at base j and assembly i at location $p_{\rm dn}(j)$. In their turn these Poisson processes are split into Poisson processes in a similar way as for assembly i at base j, and so on. Ultimately we obtain a Poisson demand process for each SKU $i \in I$ at each location $j \in J$. This is stated in the following lemma, where also the rates are given.

Lemma 7.1. *For each SKU $i \in I$ and location $j \in J$, the demand process of SKU i at location j is a Poisson process with rate $m_{i,j}$, where the rates $m_{i,j}$, $i \in I^{as}$, $j \in J^{ba}$, are given and*

$$m_{i,j} = \begin{cases} \sum_{k \in P_{mb}(i)} m_{k,j} r_{k,j} q_{k,i,j} & \text{if } i \in I \setminus I^{as},\ j \in J^{ba}; \\ \sum_{l \in C_{dn}(j)} m_{i,l}(1 - r_{i,l}) & \text{if } i \in I^{as},\ j \in J \setminus J^{ba}; \\ \sum_{k \in P_{mb}(i)} m_{k,j} r_{k,j} q_{k,i,j} & \\ + \sum_{l \in C_{dn}(j)} m_{i,l}(1 - r_{i,l}) & \text{if } i \in I \setminus I^{as},\ j \in J \setminus J^{ba}. \end{cases} \tag{7.3}$$

We now consider the inventory of a part $i \in I$ at a location $j \in J$. The inventory position is kept at a constant level $S_{i,j}$. Hence,

$$OH_{i,j}(t) + X_{i,j}(t) - BO_{i,j}(t) = S_{i,j}, \quad t \geq 0, \tag{7.4}$$

where $OH_{i,j}(t)$, $X_{i,j}(t)$, and $BO_{i,j}(t)$ denote the on-hand stock, the number of parts in repair and on order, and the number of backorders at time instant t. $X_{i,j}(t)$ is also called the *pipeline stock*. Obviously, at each time instant $t \geq 0$, either $OH_{i,j}(t) = 0$

or $BO_{i,j}(t) = 0$, or both. Therefore, if $X_{i,j}(t)$ is known, then $OH_{i,j}(t)$ and $BO_{i,j}(t)$ are known. Equation (7.4) is known as the *stock balance equation.*

Let $OH_{i,j}$, $X_{i,j}$, and $BO_{i,j}$ be random variables which denote the physical stock on hand, the number of parts in repair and on order, and the number of backorders in steady state. Then

$$OH_{i,j} + X_{i,j} - BO_{i,j} = S_{i,j}, \quad i \in I, \ j \in J.$$

By this equation the distribution of $BO_{i,j}$ and $OH_{i,j}$ can be determined from the distribution of $X_{i,j}$. For the backorder distribution, this is stated in the following lemma.

Lemma 7.2. *For all $i \in I$ and $j \in J$, the distribution of the number of backorders $BO_{i,j}$ is given by:*

$$P\{BO_{i,j} = x\} = \begin{cases} \sum_{y=0}^{S_{i,j}} P\{X_{i,j} = y\} & \text{if } x = 0; \\ P\{X_{i,j} = x + S_{i,j}\} & \text{if } x > 0. \end{cases} \tag{7.5}$$

By Lemma 7.2, we can compute backorder distributions from pipeline distributions. The other way around appears to be possible too, as we will see in the next section. There we will derive a recursive procedure for the computation of all pipeline and backorder distributions. That procedure starts with the pipeline distributions of childless SKU's at the central depot and ends with the distributions for the pipelines $X_{i,j}$ and backorders $BO_{i,}$ of all assemblies $i \in I^{\text{as}}$ at all bases $j \in J^{\text{ba}}$. Once we have the latter distributions, we can easily determine the mean waiting times $W_{i,j}(\mathbf{S})$ (use Little's law):

$$W_{i,j}(\mathbf{S}) = \frac{E\{BO_{i,j}\}}{m_{i,j}}, \quad i \in I^{\text{as}}, \ j \in J^{\text{ba}}. \tag{7.6}$$

Next, by (7.2), one finds the aggregate mean waiting times $\hat{W}_j(\mathbf{S})$, $j \in J^{\text{ba}}$.

7.3.2 Recursive Expressions for Pipelines

In this section, we describe the recursive procedure for the computation of all pipeline and backorder distributions. By Lemma 7.2, we can compute the distribution of the number of backorders $BO_{i,j}$ of a SKU i at a location j from the pipeline distribution of the same part at the same location. It is also possible to compute pipeline distributions from backorder distributions. However, the distribution of the pipeline stock $X_{i,j}$ of a SKU i at a location j is not computed from the backorder distribution of the same SKU at the same location, but from the backorder distribution of the same SKU i at the parent location $p_{\text{dn}}(j)$ (if applicable) and the backorder distributions of SKU's $k \in C_{\text{mb}}(i)$ (if applicable) at the same location j. To explain this in more detail, let $X_{i,j}^{\text{rep}}$ and $X_{i,j}^{\text{res}}$ be random variables which denote the number of parts of SKU i in repair at location j and the number of parts of SKU i on order by location j (= being resupplied to location j) in steady state. Then

$$X_{i,j} = X_{i,j}^{\text{rep}} + X_{i,j}^{\text{res}}, \quad i \in I, \ j \in J, \tag{7.7}$$

where $X_{i,j}^{\text{rep}}$ and $X_{i,j}^{\text{res}}$ are mutually independent. So, the distribution of $X_{i,j}$ can be obtained by convoluting the distributions of $X_{i,j}^{\text{rep}}$ and $X_{i,j}^{\text{res}}$. For both the *repair pipeline stocks* $X_{i,j}^{\text{rep}}$ and the *resupply pipeline stocks* $X_{i,j}^{\text{res}}$, we derive recursive expressions below. The distribution of $X_{i,j}^{\text{rep}}$ can be computed directly if SKU i is childless and from the backorder distributions of the children $k \in C_{\text{mb}}(i)$ at location j otherwise; see Lemma 7.3. The distribution of $X_{i,j}^{\text{res}}$ can be computed directly at the central depot (i.e., for $j = 0$) and from the backorder distribution of SKU i at the parent location $p_{\text{dn}}(j)$ otherwise; see Lemma 7.4. These recursive expressions, together with the expressions (7.5) and (7.7), result directly in an exact, recursive evaluation procedure. The procedure starts with the pipeline distributions of childless SKU's at the central depot and ends with pipeline and backorder distributions of assemblies at all bases; see the algorithm formulated at the end of this section.

Lemma 7.3. *Let $i \in I$ and $j \in J$.*

(i) If $i \in I^{cl}$, then the repair pipeline $X_{i,j}^{rep}$ is Poisson distributed with parameter $m_{i,j} r_{i,j} t_{i,j}^{rep}$;
(ii) If $i \in I \setminus I^{cl}$, then

$$X_{i,j}^{rep} = Y_0 + \sum_{k \in C_{mb}(i)} Y_k,$$

where Y_0 is a Poisson distributed random variable with parameter $m_{i,j} r_{i,j} t_{i,j}^{rep}$, Y_k is a random variable with

$$P\{Y_k = y\} = \sum_{x=y}^{\infty} \binom{x}{y} \left(\frac{m_{i,j} r_{i,j} q_{i,k,j}}{m_{k,j}} \right)^{y} \left(1 - \frac{m_{i,j} r_{i,j} q_{i,k,j}}{m_{k,j}} \right)^{x-y} P\{BO_{k,j} = x\}, \quad y \in \mathbb{N}_0,$$

for each $k \in C_{mb}(i)$ (with the convention that $P\{Y_k = 0\} = 1$ and $P\{Y_k = y\} = 0$ for all $y \geq 1$ when $m_{k,j} = 0$), and Y_0 and all Y_k are independent.

Proof. Let $i \in I$ and $j \in J$. Demands for SKU i at location j occur according to a Poisson process with rate $m_{i,j}$. Each demand is accompanied by a failed part that is returned. Each failed part is sent into repair with probability $r_{i,j}$. So, parts enter the repair pipeline of SKU i according to a Poisson process with rate $m_{i,j} r_{i,j}$. For each part, the repair leadtime is given by the generic random variable $T_{i,j}^{\text{rep}}$ and this time is independent of other parts being sent into repair.

If $i \in I^{\text{cl}}$, i.e., if SKU i has no children, then the repair of a failed part of SKU i cannot be delayed because some underlying part is not immediately available.

In that case we can apply *Palm's theorem* (see [10]), and we find that $X_{i,j}^{\text{rep}}$ is Poisson distributed with parameter $m_{i,j} r_{i,j} t_{i,j}^{\text{rep}}$. This completes the proof of part (i).

We now consider the case $i \in I \setminus I^{\text{cl}}$, i.e., SKU i has one or more children $k \in C_{\text{mb}}(i)$. Then each failed part has a deterministic repair leadtime $t_{i,j}^{\text{rep}} = E\{T_{i,j}^{\text{rep}}\}$, but the start of this deterministic repair leadtime may be delayed because an underlying part is required for the repair while that part is not immediately available.

Let $X_{i,j}^{\text{rep}}(t)$ denote the number of parts of SKU i in repair at location j at some time instant t (assume $t \geq t_{i,j}^{\text{rep}}$). Then

$$\begin{aligned} X_{i,j}^{\text{rep}}(t) &= \text{Parts in repair which arrived in the interval } (t - t_{i,j}^{\text{rep}}, t) \\ &+ \sum_{k \in C_{\text{mb}}(i)} [\text{Parts in repair which arrived prior to } t - t_{i,j}^{\text{rep}} \text{ and which} \\ &\qquad \text{are waiting for a backordered part of SKU } k \text{ at time instant } t - t_{i,j}^{\text{rep}}]. \end{aligned}$$

Note that the terms in the latter sum have the same distributions as the numbers of parts of SKU i waiting for a backordered part of SKU k, $k \in C_{\text{mb}}(i)$, at an arbitrary time instant. Now, let random variable Y_0 denote the number of parts of SKU i sent into repair at location j in an interval of length $t_{i,j}^{\text{rep}}$, and let Y_k, $k \in C_{\text{mb}}(i)$, denote the number of parts of SKU i which are waiting for a part of the underlying SKU k at an arbitrary instant. Then $X_{i,j}^{\text{rep}} = Y_0 + \sum_{k \in C_{\text{mb}}(i)} Y_k$. Further, Y_0 is Poisson distributed with parameter $m_{i,j} r_{i,j} t_{i,j}^{\text{rep}}$, and Y_0 is independent of the Y_k, $k \in C_{\text{mb}}(i)$, due to the non-overlapping time intervals and the fact that Poisson processes have independent increments. For the random variable Y_k we obtain a binomial distribution if we condition to the total number of backorders for SKU k at location j at an arbitrary time instant:

$$P\{Y_k = y | BO_{k,j} = x\} = \binom{x}{y} \left(\frac{m_{i,j} r_{i,j} q_{i,k,j}}{m_{k,j}} \right)^y \left(1 - \frac{m_{i,j} r_{i,j} q_{i,k,j}}{m_{k,j}} \right)^{x-y}$$

for all $0 \leq y \leq x$. The explanation for this result is as follows. Demands for SKU k at location j arrive according to a Poisson process with rate $m_{k,j}$. A fraction $m_{i,j} r_{i,j} q_{i,k,j}$ of this stream comes from parts of SKU i that need a part of SKU k for their repair at location j. Hence, each demand, and thus also each backordered request, comes from parts of SKU i being repaired at location j with probability $(m_{i,j} r_{i,j} q_{i,k,j})/m_{k,j}$. Finally, it is easily seen that the Y_k, $k \in C_{\text{mb}}(i)$, are independent of each other. This completes the proof of part (ii). □

Lemma 7.4. *Let $i \in I$ and $j \in J$.*

(i) If $j = 0$, then the resupply pipeline $X_{i,j}^{res} = X_{i,0}^{res}$ is Poisson distributed with parameter $m_{i,0}(1 - r_{i,0}) t_{i,0}^{res}$;
(ii) If $j \in J \setminus \{0\}$, then $X_{i,j}^{res} = Y_0 + Y$, where Y_0 is a Poisson distributed random variable with parameter $m_{i,j}(1 - r_{i,j}) t_{i,j}^{res}$, Y is a random variable with

$$P\{Y=y\} = \sum_{x=y}^{\infty} \binom{x}{y} \left(\frac{m_{i,j}(1-r_{i,j})}{m_{i,l}} \right)^y \left(1 - \frac{m_{i,j}(1-r_{i,j})}{m_{i,l}}\right)^{x-y} P\{BO_{i,l}=x\},\ y \in \mathbb{N}_0,$$

where $l = p_{dn}(j)$ *(with the convention that* $P\{Y=0\}=1$ *and* $P\{Y=y\}=0$ *for all* $y \geq 1$ *when* $m_{i,l}=0$*), and* Y_0 *and* Y *are mutually independent.*

Proof. The proof of this lemma is along the same lines as the proof of the preceding lemma. Note that each term of the sum in the equation in the second part of the lemma denotes the probability that out of a total of x backlogged replenishment orders at location l, y demands are from location j. □

Algorithm 7.1 *(Exact evaluation procedure)*

Step 1 Determine all $m_{i,j}$ for all i and j with $i \in I \setminus I^{\mathrm{as}}$ or $j \in J \setminus J^{\mathrm{ba}}$ by (7.3)
Step 2 For all $i = |I|, |I|-1, \ldots, 1$ and $j = 0, 1, \ldots, |J|-1$:
  Determine the distribution of $X_{i,j}^{\mathrm{rep}}$ by Lemma 7.3;
  Determine the distribution of $X_{i,j}^{\mathrm{res}}$ by Lemma 7.4;
  Determine the distribution of $X_{i,j}$ by (7.7);
  Determine the distribution of $BO_{i,j}$ by Lemma 7.2
Step 3 Determine the $W_{i,j}(\mathbf{S})$, $i \in I^{\mathrm{as}}$, $j \in J^{\mathrm{ba}}$, by (7.6);
  Determine the $\hat{W}_j(\mathbf{S})$ by (7.2)

Example 7.1 (continued). We apply the exact evaluation algorithm to evaluate the basestock policy with the following basestock levels:

$$(S_{1,0}, \ldots, S_{12,0}) = (3,3,23,12,13,9,11,6,7,12,6,9),$$
$$(S_{1,j}, \ldots, S_{12,j}) = (3,2,5,2,3,1,1,1,1,1,1,1), \quad j = 1, \ldots, 5.$$

Under this basestock policy, the total average costs are equal to $C(\mathbf{S}) = 33.25 \cdot 10^3$ Euro. By applying the exact evaluation procedure, we find the distribution the number of backorders of pump unit 1 and pump unit 2 at each base $j \in J^{\mathrm{ba}}$. We obtain the following values for $P\{BO_{1,j}=x\}$ for $x = 0, \ldots, 7$: 0.9384, 0.0414, 0.0141, 0.0044, 0.0013, 0.0003, 0.0001, 0.0000. And we find the following values for $P\{BO_{2,j}=x\}$ for $x = 0, \ldots, 5$: 0.9560, 0.0348, 0.0075, 0.0014, 0.0003, 0.0000. For the aggregate mean waiting time of an arbitrary demand at a base, we get $\hat{W}_j(\mathbf{S}) = 0.00428$ years.

7.4 Approximate Evaluation

From a computational point of view, the calculation of complete distribution functions may be less attractive. An often applied procedure then is to calculate the first two moments of backorder quantities and pipeline stocks. Below we derive recursive expressions to determine the first two moments of pipeline stocks from first two moments of backorder quantities (see Lemmas 7.6 and 7.7). For the other way around, we first have to fit a distribution function to the first two moments of a pipeline stock $X_{i,j}$, $i \in I$, $j \in J$, in order to determine the first two moments of the backorder quantity $BO_{i,j}$. This follows from the expressions in Lemma 7.5 below. Lemma 7.5 follows directly from Lemma 7.2. Lemmas 7.6 and 7.7 follow after some tedious but straightforward calculations based on the expressions given in Lemmas 7.3 and 7.4 (see also the derivation of Eqs. (6.20) and (6.21) in Sect. 6.3.2).

Lemma 7.5. *For all $i \in I$ and $j \in J$, the first two moments of the number of backorders $BO_{i,j}$ are given by:*

$$E\{BO_{i,j}\} = E\{X_{i,j}\} - S_{i,j} + \sum_{x=0}^{S_{i,j}} (S_{i,j} - x)P\{X_{i,j} = x\},$$

$$E\{BO_{i,j}^2\} = E\{X_{i,j}^2\} - 2S_{i,j}E\{X_{i,j}\} + S_{i,j}^2 - \sum_{x=0}^{S_{i,j}} (S_{i,j} - x)^2 P\{X_{i,j} = x\}.$$

Lemma 7.6. *Let $i \in I$ and $j \in J$.*

(i) If $i \in I^{cl}$, then

$$E\{X_{i,j}^{rep}\} = Var\{X_{i,j}^{rep}\} = m_{i,j} r_{i,j} t_{i,j}^{rep};$$

(ii) If $i \in I \setminus I^{cl}$, then

$$E\{X_{i,j}^{rep}\} = m_{i,j} r_{i,j} t_{i,j}^{rep} + \sum_{k \in C_{mb}(i)} f_k E\{BO_{k,j}\},$$

$$Var\{X_{i,j}^{rep}\} = m_{i,j} r_{i,j} t_{i,j}^{rep} + \sum_{k \in C_{mb}(i)} \left[f_k(1 - f_k)E\{BO_{k,j}\} + f_k^2 Var\{BO_{k,j}\} \right],$$

where $f_k := (m_{i,j} r_{i,j} q_{i,k,j})/m_{k,j}$ for all $k \in C_{mb}(i)$.

Lemma 7.7. *Let $i \in I$ and $j \in J$.*

(i) If $j = 0$, then $X_{i,j}^{res} = X_{i,0}^{res}$ and

$$E\{X_{i,0}^{res}\} = Var\{X_{i,0}^{res}\} = m_{i,0}(1 - r_{i,0}) t_{i,0}^{res};$$

(ii) If $j \in J \setminus \{0\}$, then

$$E\{X_{i,j}^{res}\} = m_{i,j}(1 - r_{i,j}) t_{i,j}^{res} + f E\{BO_{i,l}\},$$

$$Var\{X_{i,j}^{res}\} = m_{i,j}(1 - r_{i,j}) t_{i,j}^{res} + f(1 - f)E\{BO_{i,l}\} + f^2 Var\{BO_{i,l}\},$$

where $l = p_{dn}(j)$ and $f := (m_{i,j}(1 - r_{i,j}))/m_{i,l}$.

In the approximate evaluation procedure, we go through the combinations of SKU's $i \in I$ and locations $j \in J$ in the same order as in the exact evaluation procedure (Algorithm 7.1). Per step, we compute the mean and variance of $X_{i,j}^{\text{rep}}$, $X_{i,j}^{\text{res}}$, and $X_{i,j}$ via the above results, we fit a distribution on $X_{i,j}$, and we compute the mean and variance of $BO_{i,j}$. For the fit of a distribution on $X_{i,j}$, we use the two-moment fit as described in Appendix "Two-Moment Fit of Adan et al. [1]" of Chap. 6 (cf. Adan et al. [1]). The approximate evaluation is as follows.

Algorithm 7.2 *(Approximate evaluation based on two-moment fits)*

Step 1 Determine all $m_{i,j}$ for all i and j with $i \in I \setminus I^{\text{as}}$ or $j \in J \setminus J^{\text{ba}}$ by (7.3)
Step 2 For all $i = |I|, |I|-1, \ldots, 1$ and $j = 0, 1, \ldots, |J|-1$:
  Determine the mean and variance of $X_{i,j}^{\text{rep}}$ by Lemma 7.6;
  Determine the mean and variance of $X_{i,j}^{\text{res}}$ by Lemma 7.7;
  Determine the mean and variance of $X_{i,j}$ by (7.7);
  Apply a two-moment fit for $X_{i,j}$ (cf. Appendix "Two-Moment Fit of Adan et al. [1]" of Chap. 6);
  Determine the mean and variance of $BO_{i,j}$ by Lemma 7.5
Step 3 Determine the $W_{i,j}(\mathbf{S})$, $i \in I^{\text{as}}$, $j \in J^{\text{ba}}$, by (7.6);
  Determine the $\hat{W}_j(\mathbf{S})$ by (7.2)

Remark 7.1. Suppose that we allow that order and ship times and repair leadtimes for all SKU's and all locations are stochastic. Then the exact evaluation is not exact anymore. It still makes sense to apply the approximate evaluation. In that procedure we then only use the means of all order and ship times and repair leadtimes. It has been shown that the approximate evaluation procedure is still accurate in that case; see Avşar and Zijm [3].

7.5 Greedy Heuristic

The greedy heuristic is similar as in previous chapters. We divide the set $\mathscr{S}$ of all solutions into a subset

$$\mathscr{S}^{\text{feas}} := \{\mathbf{S} \in \mathscr{S} | \hat{W}_j(\mathbf{S}) \leq \hat{W}_j^{\text{obj}} \text{ for all } j \in J^{\text{ba}}\}$$

of feasible solutions and the subset $\mathscr{S} \setminus \mathscr{S}^{\text{feas}}$ of non-feasible solutions. Next, for each solution $\mathbf{S} \in \mathscr{S}$, we define the distance $d(\mathbf{S})$ to the set $\mathscr{S}^{\text{feas}}$ of feasible solutions as

$$\begin{aligned} d(\mathbf{S}) &:= \sum_{j \in J^{\text{ba}}} \left(\hat{W}_j(\mathbf{S}) - \hat{W}_j^{\text{obj}} \right)^+ \\ &= \sum_{j \in J^{\text{ba}}} \left(\sum_{i \in I^{\text{as}}} \frac{m_{i,j}}{M_j} W_{i,j}(\mathbf{S}) - \hat{W}_j^{\text{obj}} \right)^+ . \end{aligned}$$

In each greedy step, we look at the ratio of the decrease in distance to the set of feasible solutions and the increase in costs if one of the basestock levels $S_{i,j}$ is increased by one unit. Let $i \in I$ and $j \in J$, and suppose that $S_{i,j}$ would be increased by 1 unit. Then the increase in costs is equal to (use (7.1))

$$\Delta_{i,j}C(\mathbf{S}) = \Delta_j C_i(\mathbf{S}_i) = C_i(\mathbf{S}_i + \mathbf{e}_j) - C_i(\mathbf{S}_i) = c_i^{\mathrm{h}},$$

where $\mathbf{e}_j$ is a $|J|$-dimensional unit vector with positions $0, 1, \ldots, |J|-1$ and a 1 on position j. The decrease in distance to the feasible solutions is given by $-\Delta_{i,j}d(\mathbf{S})$. For each $i \in I$ and $j \in J$, let $\mathbf{E}_{i,j}$ be an $|I| \times |J|$ matrix with positions (i', j'), $i' \in I$, $j' \in J$, a 1 on the (i, j)-th position and zeros on all other positions. We obtain

$$\Delta_{i,j}d(\mathbf{S}) = d(\mathbf{S} + \mathbf{E}_{i,j}) - d(\mathbf{S}). \tag{7.8}$$

This time, we have no further simplification for $\Delta_{i,j}d(\mathbf{S})$ in general because of the general multi-indenture structure. It does hold however, that an increase of $S_{i,j}$ for some $i \in I$ and $j \in J$ has only an effect on pipeline and backorder distributions for combinations (i', j'), $i' \in I$, $j' \in J$, for which:

- $i' = i$ or i' is a direct or indirect parent of i; and,
- $j' = j$ or j' is a direct or indirect child of j.

The factor

$$\Gamma_{i,j} = \frac{-\Delta_{i,j}d(\mathbf{S})}{\Delta_{i,j}C(\mathbf{S})} = \frac{-\Delta_{i,j}d(\mathbf{S})}{c_i^{\mathrm{h}}} \tag{7.9}$$

denotes the decrease in distance to the feasible solutions per unit of increase in costs. For the combination of SKU i and local warehouse j with the highest ratio $\Gamma_{i,j}$, we have the "biggest bang for the buck" and we increase the basestock level to $S_{i,j}+1$. We continue with the above greedy steps until we arrive at a feasible solution. The formal procedure is described in Algorithm 7.3.

Algorithm 7.3 (Greedy heuristic)

Step 1 Set $S_{i,j} := 0$, $i \in I$, $j \in J$

Step 2

Step 2-a Calculate $\Delta_{i,j}d(\mathbf{S})$, and $\Gamma_{i,j}$, $i \in I$, $j \in J$ (cf. (7.8) and (7.9))

Step 2-b While $d(\mathbf{S}) > 0$:

(a) Determine i' and j' such that $\Gamma_{i',j'} \geq \Gamma_{i,j}$, $i \in I$, $j \in J$;

(b) Set $S_{i',j'} := S_{i',j'} + 1$;

(c) Calculate $\Delta_{i,j}d(\mathbf{S})$, and $\Gamma_{i,j}$, $i \in I$, $j \in J$ (cf. (7.8)–(7.9))

7.6 Case: Royal Netherlands Navy

In 2000, Rustenburg [11] applied the multi-echelon, multi-indenture model for a case at the Royal Netherlands Navy (RNN). For a few technical systems, he investigated existing total stocks of spare parts and the corresponding system availabilities, and he compared them to alternative, smarter investments in spare parts and their corresponding system availabilities. These smarter investments were calculated by a greedy heuristic that has been tailored for the RNN situation. This heuristic generates good combinations for the total investment in spare parts and system availability and is described in Sect. 7.6.1. Next, in Sect. 7.6.2, we give the case results for two systems: a goalkeeper and a radar system. For more results and details, we refer to [11].

7.6.1 Optimizing Inventory Investment and System Availability

At the RNN, one has a distribution network consisting of a central depot and a number of bases. These bases are frigates and other ships at which technical systems are installed. At these ships, one has certain repair facilities and a stockroom for spare parts on board. The RNN is interested in the balance between the total investment in spare parts and system availability. Below we describe the problem formulation and corresponding greedy heuristic that one obtains for the combination of these two criteria.

Let $\hat{C}_i(\mathbf{S}_i)$ be the investment in spare parts of SKU i, $i \in I$. It holds that

$$\hat{C}_i(\mathbf{S}_i) = c_i^{\mathrm{a}} \sum_{j \in J} S_{i,j},$$

where c_i^{a} is the price of a part of SKU i. For a basestock policy $\mathbf{S}$, the total investment is equal to

$$\hat{C}(\mathbf{S}) = \sum_{i \in I} \hat{C}_i(\mathbf{S}_i). \tag{7.10}$$

For the technical systems that we consider in Sect. 7.6.2, it holds that one has one unit of them on board of each frigate. Hence, having a backorder for one of the assemblies at a frigate implies that the system is down. Hence, for the system availability of the unit at base j, we have the following expression (see also Sect. 2.7.3):

$$A_j(\mathbf{S}) = \prod_{i \in I^{\mathrm{as}}} (1 - P\{BO_{i,j} > 0\}), \quad j \in J^{\mathrm{ba}}, \tag{7.11}$$

and the average system availability equals

$$A(\mathbf{S}) = \frac{1}{|J^{\mathrm{ba}}|} \sum_{j \in J^{\mathrm{ba}}} A_j(\mathbf{S}). \tag{7.12}$$

The optimization problem is then formulated as:

$$\begin{array}{lll} \text{(Q)} & \min & \hat{C}(\mathbf{S}) \\ & \max & A(\mathbf{S}) \\ & \multicolumn{2}{l}{\text{subject to } \mathbf{S} \in \mathscr{S}.} \end{array}$$

We are mainly interested in basestock policies $\mathbf{S}$ under which the system availabilities $A_j(\mathbf{S})$ are relatively high (above 0.8, say). Then these system availabilities are well approximated by

$$A_j(\mathbf{S}) \approx 1 - \sum_{i \in I^{\mathrm{as}}} P\{BO_{i,j} > 0\}, \quad j \in J^{\mathrm{ba}}.$$

For the average system availability, it then holds that

$$\begin{aligned} A(\mathbf{S}) &\approx 1 - \frac{1}{|J^{\mathrm{ba}}|} \sum_{j \in J^{\mathrm{ba}}} \sum_{i \in I^{\mathrm{as}}} P\{BO_{i,j} > 0\} \\ &= 1 - \frac{1}{|J^{\mathrm{ba}}|} PBO(\mathbf{S}), \end{aligned}$$

where

$$PBO(\mathbf{S}) := \sum_{j \in J^{\mathrm{ba}}} \sum_{i \in I^{\mathrm{as}}} P\{BO_{i,j} > 0\}. \tag{7.13}$$

This shows that maximizing $A(\mathbf{S})$ is roughly similar to minimizing $PBO(\mathbf{S})$. Hence, instead of solving Problem (Q), we may also consider the following problem:

$$\begin{array}{lll} \text{(Q')} & \min & \hat{C}(\mathbf{S}) \\ & \min & PBO(\mathbf{S}) \\ & \multicolumn{2}{l}{\text{subject to } \mathbf{S} \in \mathscr{S}.} \end{array}$$

For Problem (Q'), we have two objective functions consisting of sums over SKU's i and locations j, and we can apply the logic of greedy procedures. The function $\hat{C}(\mathbf{S})$ is linear. The function $PBO(\mathbf{S})$ consists of terms $P\{BO_{i,j} > 0\}, i \in I^{\mathrm{as}}$, $j \in J^{\mathrm{ba}}$, which are decreasing as a function of each of the basestock levels $S_{i,j}$. Generally, they are not convex for all $S_{i,j} \geq 0$, and hence one has to start the greedy algorithm with values that may be larger than 0. For the problems studied in Chap. 2, we were able to derive analytically the conditions under which convexity is guaranteed. Here, one can find an appropriate starting solution by computational experiments (cf. [11]):

$$S_{i,j} := round(m_{i,j}(r_{i,j}t_{i,j}^{\mathrm{rep}} + (1 - r_{i,j})t_{i,j}^{\mathrm{res}})), \quad i \in I, \ j \in J^{\mathrm{ba}}, \tag{7.14}$$

$$S_{i,0} := round(\frac{1}{2}m_{i,0}(r_{i,0}t_{i,0}^{\text{rep}} + (1 - r_{i,j})t_{i,0}^{\text{res}})), \quad i \in I, \tag{7.15}$$

where $round(x)$ stands for the standard rounded value for $x \in \mathbb{R}$. For the greedy steps, for each $i \in I$ and $j \in J$, one looks at the ratio $\Gamma_{i,j}$ of the decrease in $PBO(\mathbf{S})$ and the increase in $\hat{C}(\mathbf{S})$. It holds that

$$\begin{aligned} \Delta_{i,j}\hat{C}(\mathbf{S}) &= \Delta_j\hat{C}_i(\mathbf{S}_i) = \hat{C}_i(\mathbf{S}_i + \mathbf{e}_j) - \hat{C}_i(\mathbf{S}_i) = c_i^{\text{a}}, \\ \Delta_{i,j}PBO(\mathbf{S}) &= PBO(\mathbf{S} + \mathbf{E}_{i,j}) - PBO(\mathbf{S}), \end{aligned}$$

and thus the ratio $\Gamma_{i,j}$ is equal to

$$\Gamma_{i,j} = \frac{-\Delta_{i,j}PBO(\mathbf{S})}{\Delta_{i,j}\hat{C}(\mathbf{S})} = \frac{-\Delta_{i,j}PBO(\mathbf{S})}{c_i^{\text{a}}}. \tag{7.16}$$

We obtain the following greedy algorithm which generates a set $\mathscr{E}$ of close-to-efficient solutions for Problem (Q')

Algorithm 7.4 *(Greedy algorithm for Problem (Q'))*

Step 1 Set $S_{i,j}$, $i \in I$, $j \in J$ cf. (7.14) and (7.15);
$\mathscr{E} := \{\mathbf{S}\}$;
Calculate $\hat{C}(\mathbf{S})$ and $PBO(\mathbf{S})$ (by (7.10) and (7.13)).
Step 2 Calculate $\Gamma_{i,j}$, $i \in I$, $j \in J$ (by (7.16));
Determine i' and j' such that $\Gamma_{i',j'} \geq \Gamma_{i,j}$, $i \in I$, $j \in J$;
$\mathbf{S} := \mathbf{S} + \mathbf{E}_{i',j'}$;
$\mathscr{E} := \mathscr{E} \cup \{\mathbf{S}\}$.
Step 3 Calculate $\hat{C}(\mathbf{S})$ and $PBO(\mathbf{S})$ (by (7.10) and (7.13));
If 'stop criterium', then stop, else goto Step 2.

In Algorithm 7.4, while generating close-to-efficient solutions $\mathbf{S}$ for Problem (Q'), one can compute the corresponding average system availabilities $A(\mathbf{S})$ (by (7.11) and (7.12)). Generally, the close-to-efficient solutions $\mathbf{S}$ for Problem (Q') are also close-to-efficient solutions for Problem (Q).

7.6.2 Case Results

In this section, we present results for two technical system at the RNN. The first system is the *Goalkeeper*, a close-in weapon system primarily designed to intercept fast and low incoming missiles, as well as aircrafts, helicopters and surface targets. In the RNN, we find these systems on frigates as well as auxiliary ships. All Goalkeepers are more or less identical. Hence we consider all ships on which such a system is installed. The second system is the *long range air surveillance radar*, type LW-08/02, which is installed on Multi-Purpose frigates (M-frigates). The primary

Fig. 7.4 A photo of the long range air surveillance radar, type LW-08/02

function of the LW-08/02 is the timely detection of air targets. For that purpose the LW-08/02 is equipped with a long distance radar with a small minimum range and a high resistance to electronic counter measures. A picture of the LW-08/02 is shown in Fig. 7.4.

For our studies we selected the SKU's with a price $\geq$ 34.0 Euros (for the original results in [15], all results were in Netherlands Guilders (NLG); here, we recalculated them for Euros, with conversion rate 1 NLG = 0.4538 Euro). Moreover the SKU's with a registered demand rate equal to zero are excluded. With this selection, we reduced the number of SKU's considered in the cases by approximately 12 %, while the inventory investment considered was approximately 98 % of the complete investment in spare parts of the considered systems. For the total number of spare parts with corresponding inventory investment and the selected spare parts with corresponding inventory investment, see Table 7.3.

In order to judge the actual investments properly (recall that actual refers to the actual investment in the year 2000), we should note that the RNN Maintenance Company used an inventory control policy that can be classified as an item approach (cf. Sect. 2.6). Their method aimed to achieve target fill rates, that were equal for most SKU's, i.e., independent of their level in the material breakdown structure as well as their price. The application of the greedy heuristic (Algorithm 7.4) leads to

Table 7.3 Selected number of spare parts for the two technical systems

System	Total no. of SKU's	Inventory investment (in 10^6 Euro)	Selected nr of parts	Inv. investm. selected SKU's (in 10^6 Euro)
Goalkeeper	771	51.7	675	50.8
LW-08/02	140	12.0	133	11.8

Table 7.4 Current situation versus an efficient situation

	Goalkeeper		LW-08/02	
	actual	efficient	actual	efficient
Inventory investment (in 10^6 Euro)	50.8	46.3	11.8	4.7
availability (%)	56.3	90.3	88.9	90.7

stock allocations in which an efficient balance is found between investments (and thus implicitly fill rates) of cheap and expensive SKU's and of higher and lower level SKU's in the material breakdown structure. In addition, a balance is found between stocks at the central depot and the bases (ships).

In Table 7.4, we compare the current situation with the performance under a close-to-efficient solution, as obtained by Algorithm 7.4 (in combination with the approximate evaluation procedure). The selected efficient solution is the first generated solution with an average system availability of at least 90 %. Its availability and the corresponding inventory investment are denoted as an "efficient situation". We learn from Table 7.4 that for both technical systems the actual investments in spare parts and availabilities were far from being efficient. When studying the Goalkeeper, we observe that the inventory investment in the efficient situation was relatively close to the actual inventory investment. However, the availability in the actual situation was much lower than in the efficient situation. Hence, for the Goalkeeper, one had a substantial *mis*investment in spare parts. The actual availability of the LW-08/02 was rather close to the 90 % availability of the efficient situation. However, the actual investment was much higher than the investment of the efficient situation. So, here one had to do with a substantial *over*investment.

7.7 Concluding Remarks

The multi-echelon, multi-indenture model of this chapter generalizes Sherbrooke's METRIC model [13] (two echelon levels, single indenture level) that we presented in Chap. 6. For literature on extensions of the METRIC model to systems with more than two echelon levels, see Sect. 6.6. Here, we focus on discussing relevant literature on multi-indenture systems.

The existence of two indenture levels was introduced by Sherbrooke [14] for a single-location inventory problem, and he derived an approximate evaluation method. Muckstadt [8] was the first to consider a two-echelon, two-indenture system, and he developed an approximate evaluation method that was similar to the METRIC approach for the METRIC model (i.e., based on single-moment fits). Sherbrooke [15] considered the same system and described an approximate evaluation method based on two-moment fits. He extended the approach of Slay [17] for the METRIC model. The approximate method that we presented in Sect. 7.4 for a multi-echelon, multi-indenture system is in the same line (our approach follows the method of Graves [7]). Rustenburg et al. [12] considered a multi-echelon, multi-indenture system and developed an exact evaluation method based on recursive expressions for pipeline stocks and numbers of backorders. This method has been described in Sect. 7.3 and generalizes the exact method of Graves for the METRIC model. For the heuristic optimization of the basestock levels, various authors used greedy heuristics; see [9, 12, 15, 16].

Some further extensions have been considered as well. For multi-echelon, multi-indenture systems with capacitated repair shops, see [18, 20], and for the combination with the decisions on where to repair failed parts and the investment in spare parts equipment, see [5].

While for the above models, it is assumed that at most one subassembly is needed for the repair of an assembly, Van Jaarsveld et al. [19] and Driessen et al. [6] observed that in several repair shops in practice multiple subassemblies are needed for the repair of one assembly and one only knows which subassemblies are needed after an investigation of the cause of the failure of the assembly. In addition, the potential subassemblies that might be needed can come from a wide subset of subassemblies. Driessen et al. give empirical evidence that this complicating feature leads to serious problems for the control of repair shops. Given this complicating feature, Van Jaarsveld et al. develop a method for the inventory control of the subassemblies (they focus on the repair leadtimes for assemblies and preventing delays because of a lack of subassemblies), where they make use of the similarity with the inventory control problem for assemble-to-order systems.

Problems

7.1. We consider a specific problem instance for the model studied in this chapter. Suppose that we have only one location, i.e., $J = J^{\text{ba}} = \{0\}$. Further, we assume that there is only one assembly and one subassembly. These components are denoted as SKU 1 and SKU 2, respectively; i.e., $|I| = 2$, and $I^{\text{as}} = \{1\}$. The demand rate for the assembly is 1 demand per month. It is always possible to repair a failed assembly. The failure of an assembly is due to a failure of the subassembly with probability 0.6, and otherwise it is due to an underlying part of which one always has ample stock available. The repair leadtime for the assembly is 0.5 months (this repair leadtime is deterministic). The subassembly is a consumable, i.e., each failed subassembly is disposed of and replaced by a new assembly which is ordered at an external supplier. The corresponding procurement leadtime is equal to 1 month. The basestock level for both the assembly and the subassembly is 2.

Determine the mean waiting time for the assembly, $W_{1,0}(\mathbf{S})$, via the exact evaluation procedure.

References

1. Adan, I.J.B.F., Van Eenige, M.J.A., Resing, J.A.C.: Fitting discrete distributions on the first two moments. Probab. Eng. Inf. Sci. **9**, 623–632 (1996)
2. Alfredsson, P.: Optimization of multi-echelon repairable item inventory systems with simultaneous location of repair facilities. Eur. J. Oper. Res. **99**, 584–595 (1997)
3. Avşar, Z.M., Zijm, W.H.M.: Capacitated two-echelon inventory models for repairable item systems. In: Gershwin, S., et al. (eds.) Analysis and Modeling of Manufacturing Systems. Kluwer Academic, Boston (2002)
4. Basten, R.J.I., Van der Heijden, M.C., Schutten, J.M.J.: Joint optimization of level of repair analysis and spare parts stocks. Eur. J. Oper. Res. **222**, 474–483 (2012)
5. Basten, R.J.I., Van der Heijden, M.C., Schutten, J.M.J., Kutanoglu, E.: An approximate approach for the joint problem of level of repair analysis and spare parts stocking. Ann. Oper. Res. **224**, 121–145 (2015)
6. Driessen, M.A., Wiers, V.C.S., Van Houtum, G.J., Rustenburg, W.D.: Typology of repair shops for maintenance spare parts. Beta Research School WP-427, Eindhoven University of Technology (2013)
7. Graves, S.C.: A multi-echelon inventory model for a repairable item with one-for-one replenishment. Manag. Sci. **31**, 1247–1256 (1985)
8. Muckstadt, J.: A model for a multi-item, multi-echelon, multi-indenture inventory system. Manag. Sci. **20**, 472–481 (1973)
9. Muckstadt, J.A.: Analysis and Algorithms for Service Parts Supply Chains. Springer, New York (2005)
10. Palm, C.: Analysis of the Erlang traffic formula for busy-signal arrangements. Ericsson Tech. **4**, 204–212 (1938)
11. Rustenburg, W.D.: A system approach to budget-constrained spare parts management. Ph.D. thesis, Eindhoven University of Technology (2000). http://w3.tue.nl/en/services/library/digilib/publications_from_tue/dissertations/
12. Rustenburg, W.D., Van Houtum, G.J., Zijm, W.H.M.: Exact and approximate analysis of multi-echelon, multi-indenture spare parts systems with commonality. In: Shanthikumar, J.G., Yao, D.D., Zijm, W.H.M. (eds.) Stochastic Modeling and Optimization of Manufacturing Systems and Supply Chains, pp. 143–176. Kluwer Academic, Boston/Dordrecht/London (2003)[1]
13. Sherbrooke, C.C.: METRIC: a multi-echelon technique for recoverable item control. Oper. Res. **16**, 122–141 (1968)
14. Sherbrooke, C.C.: An evaluator of operational ready aircraft in a multilevel supply system. Oper. Res. **19**, 618–635 (1971)
15. Sherbrooke, C.C.: VARI-METRIC: improved approximations for multi-indenture, multi-echelon availability models. Oper. Res. **34**, 311–319 (1986)
16. Sherbrooke, C.C.: Optimal Inventory Modeling of Systems: Multi-echelon Techniques. Kluwer Academic, Boston/Dordrecht/London (2004)

[1] The material of this book chapter has been reused in this chapter with kind permission from Springer Science+Business Media.

17. Slay, F.M.: VARI-METRIC: an approach to modeling multi-echelon resupply when the demand process is Poisson with a Gamma prior. Report AF301-3, Logistics Management Institute, Washington, DC (1984)
18. Sleptchenko, A., Van der Heijden, M.C., Van Harten, A.: Effects of finite repair capacity in multi-echelon, multi-indenture service part supply systems. Int. J. Prod. Econ. **79**, 209–230 (2002)
19. Van Jaarsveld, W., Dollevoet, T., Dekker, R.: Spare parts inventory control for an aircraft component repair shop. Working paper, Erasmus University (2014)
20. Zijm, W.H.M., Avşar, Z.M.: Capacitated two-indenture models for repairable item systems. Int. J. Prod. Econ. **81–82**, 573–588 (2003)

Chapter 8
Static Repair Priorities

8.1 Introduction

In all models studied until now, the (mean) repair and procurement leadtimes are given. In practice, these leadtimes are set in consultation with suppliers, e.g. as part of coordination agreements with internal repair shops, or in repair/purchase contracts with external suppliers. These leadtimes are very important because they have a direct, strong effect on the total spare parts investment needed to meet certain system-oriented service targets. E.g., in several real-life networks, one has a central warehouse and many local warehouses with replenishment leadtimes of 1–2 weeks for the local warehouses and repair and procurement leadtimes of 3–12 months at the central warehouse, and the total stock is then to a large extent determined by these leadtimes at the central warehouse. The repair and procurement leadtimes form another dimension where one can work on higher system-oriented service levels against lower total costs. One can consider whether all leadtimes can be decreased, possibly against slightly higher unit repair/procurement prices. This has been studied by Van der Heijden et al. [17]; this study includes a case at Thales, for which a spare parts costs reduction of 5.6 % has been demonstrated. Another option is to look at setting priorities in repair shops for certain SKU's. This option is considered in this chapter.

The situation considered in this chapter is as follows. We look at a single repair shop and all repairables (SKU's) that are repaired in that repair shop. We assume that these repairables are used for a group of machines that all belong to the same machine type. Further, we assume that these machines are installed in one region and that all ready-for-use parts are kept on stock in a single location. We have no condemnation, and thus the inventory position of each repairable is kept at a constant level, i.e., we can say that we have basestock control. The standard way of working for this repair shop would be to repair all failed parts in a First-Come, First-Served

G.J. van Houtum, B. Kranenburg, *Spare Parts Inventory Control under System Availability Constraints*, International Series in Operations Research & Management Science 227, DOI 10.1007/978-1-4899-7609-3_8

(FCFS) order. We use this service discipline as our benchmark. Given the FCFS service discipline, we can use the basestock levels, i.e., the initial amounts of repairables that are taken on stock, to obtain an optimal balance between the costs for the spare parts inventory and a system-oriented service measure. Because this is done via a multi-item model, this approach classifies as a system approach.

What happens under a system approach for the optimization of the basestock levels, is that relatively low basestock levels are chosen for expensive repairables and this is compensated by relatively high basestock levels for cheaper repairables. One then still meets a system-oriented service target, but the total spare parts costs become lower. A similar idea can be used in repair shops for the order in which failed parts are treated. One can give a higher priority to expensive repairables, so that their repair leadtimes are reduced, and thus their basestock levels can become lower than under the FCFS discipline. At the same time, repair leadtimes of cheaper repairables become longer and their base stock levels have to be increased. The net result may be that the total costs are reduced, while still meeting the system-oriented service target. The main question is by how much total costs can be reduced via setting priorities in a repair shop.

We will investigate the main question on the basis of a simplified model. We model the repair shop as a single exponential server, i.e., we assume that there is a single repairman who needs exponentially distributed times for the repair of failed parts. Further, we consider *static priorities*, under which all repairables are assigned to a given number of priority classes. The single server always works on failed parts belonging to the highest priority class (within a priority class parts are treated in a FCFS order). We further assume that all repairables have the same mean repair time. Under these assumptions, we are able to find the optimal policy under the FCFS approach for instances of an arbitrary size. Further, we are able to find optimal policies under the static priority concept for small instances (low number of repairables and 2–3 priority classes), and we will develop heuristics which perform well for large instances. At the end of the chapter, we will apply our model to stylized cases, in which we feed our model with data from 4 repair shops of the Royal Netherlands Navy (we call these cases stylized because we work with a strongly simplified model). We use actual data on all SKU's that are handled by one repair shop and on prices and demand rates of these SKU's. We find in these cases costs reductions of more than 40 % when comparing the static priorities control concepts to the FCFS concept (recall that we follow already a system approach under the FCFS concept; so, these reductions are extra reductions that one can obtain!). These high costs reductions indicate that there is a significant potential for further performance improvements in real-life systems via setting priorities in repair shops.

Notice that simplifying assumptions are common in the field of capacitated production/inventory systems (see e.g. Buzacott and Shanthikumar [2] and Hopp and Spearman [6]). Evaluation and optimization methods for these simplified models constitute a basis for the analysis of more complicated systems. And, insights on the performance under different control concepts, often directly extend to more complicated systems.

The organization of this chapter is as follows. The model is described in Sect. 8.2. Next, we discuss the exact evaluation of a given solution in Sect. 8.3 and we discuss exact and heuristic optimization approaches in Sect. 8.4. After that, we investigate the performance of the heuristic approaches via computational experiments in Sect. 8.5. In that section, we also compare FCFS and static priorities for a theoretical test bed. In Sect. 8.6, the same comparison is made for the stylized cases at the Royal Netherlands Navy. Finally, concluding remarks are given in Sect. 8.7.

This chapter is largely based on Adan et al. [1].

8.2 Model

We consider a single location with one repair shop and one stockpoint, where spare parts of multiple, critical repairables are kept on stock to serve an installed base of machines in a certain region. All machines are of the same machine type. We have a set I of repairables, which are also denoted as SKU's. These SKU's are numbered $1,\ldots,|I|$ ($|I| \in \mathbb{N}$), and they occur in the configurations of the machines and are subject to failures. When a part fails in a machine, immediately a demand is placed for a spare part of the same SKU at the stockpoint and the failed part is sent to the repair shop. If the requested spare part is on stock, then the demand is immediately fulfilled. Otherwise the demand is backordered and fulfilled as soon as a spare part of the requested SKU becomes available from the repair shop. In the latter case, the machine with the part failed is down until the spare part is delivered, which, of course, is a situation that we like to prevent.

The lifetime of the machines is assumed to be long (e.g., 20–30 years). We therefore formulate a model with an infinite time horizon and policies for the inventory control and repair order at the repair shop are analyzed via a steady-state analysis. Time is continuous and the time horizon is $[0,\infty)$.

We assume that, for the total installed base, failures of an SKU occur according to a Poisson process with a constant rate. The total failure rate for SKU i is given by m_i (> 0). The assumption of a Poisson failure process is justified if lifetimes of parts are simply exponential or if the set of machines is that large that the merged stream of failure processes of individual technical systems is close to Poisson. The assumption of constant failure rates is justified if the installed base is sufficiently large or if the installed base is not that large but downtimes of systems are relatively short (e.g., because of a high service level at the stockpoint).

Let S_i ($\in \mathbb{N}_0 := \mathbb{N} \cup \{0\}$) be the number of spare parts of SKU i that is taken on stock at time instant $t = 0$. We assume that all failed parts can always be repaired. This implies that the inventory position of SKU i is always equal to S_i. I.e., at any time instant $t \geq 0$, it holds that

$$OH_i(t) + X_i(t) - BO_i(t) = S_i, \quad i \in I, \tag{8.1}$$

where $OH_i(t)$, $X_i(t)$, and $BO_i(t)$ denote the on-hand stock, the number of parts in repair, and the number of backorders at time instant t. We call the amount S_i the *spare parts stock* of SKU i, and we also refer to S_i as the *basestock level* for SKU i. Obviously, at each time instant t, either $OH_i(t) = 0$ or $BO_i(t) = 0$, or both. Therefore, if $X_i(t)$ is known, then $OH_i(t)$ and $BO_i(t)$ are known: $OH_i(t) = (S_i - X_i(t))^+$ and $BO_i(t) = (X_i(t) - S_i)^+$, where $x^+ = \max\{0, x\}$ for any $x \in \mathbb{R}$. Let OH_i, X_i, and BO_i be random variables which denote the on-hand stock, the number of parts in repair, and the number of backorders in steady state. Then,

$$OH_i + X_i - BO_i = S_i, \quad i \in I, \tag{8.2}$$

and $OH_i = (S_i - X_i)^+$ and $BO_i = (X_i - S_i)^+$.

The repair shop is modeled as a single exponential server, which repairs the failed parts of all SKU's. All repair times are exponentially distributed and mutually independent. For all SKU's, we assume the same service rate μ. To obtain a stable system, we assume $\mu > \sum_{i \in I} m_i$. The SKU's are divided over a set K of priority classes ($|K| \in \mathbb{N}$). These classes are numbered $1, \ldots, |K|$. The smaller the class number, the higher its priority. Further, we assume *preemption*, i.e., the repair of a failed part is interrupted as soon as a failed part of a higher priority class arrives at the repair shop. The assignment of SKU's to priority classes is described by a matrix $\mathbf{U}$ of size $|I| \times |K|$. The (i,k)-th element of this matrix is denoted by $u_{i,k}$:

$$u_{i,k} = \begin{cases} 1 & \text{if SKU } i \text{ is assigned to priority class } k; \\ 0 & \text{otherwise} \end{cases}$$

We also use the notation k_i to denote the class k to which SKU i has been assigned; i.e., k_i is the unique class index k for which $u_{i,k} = 1$. For each priority class k, the total stream of arriving failed parts is a Poisson process that is constituted by Poisson arrival streams of all SKU's i that are assigned to class k (i.e., for which $u_{i,k} = 1$). The total arrival rate for class k equals $M_k = \sum_{i \in I} u_{i,k} m_i$. We do not require that each class gets at least one SKU assigned. Hence, some classes may be empty, and have a zero arrival rate.

Under the above assumptions, *the control and behavior of the repair shop is independent of actual on-hand stocks and the basestock levels* S_i. The repair shop has a steady-state behavior that is equivalent to the behavior of a priority queueing system. In general, for these priority systems, it is hard to derive steady-state distributions for the numbers of jobs of the various types. Our assumptions with respect to the repair shop and the repairs of failed parts are chosen such that an efficient evaluation of these steady-state distributions is possible. Obviously, these assumptions do not reflect reality, but they do capture the essential feature that different SKU's compete for the same repair capacity. Recall that these assumptions are common in the field of capacitated production/inventory systems (see e.g. [2] and [6]).

For the costs, we distinguish inventory holding costs and penalty costs. The inventory holding costs consist of costs for invested capital in spare parts and storage costs. The investment costs are made at time $t = 0$ and are linear in S_i. These costs may be translated into a linear cost that is paid per spare part per time unit.

The storage costs are paid per time unit and they are linear in the on-hand stock. We assume that the costs for invested capital dominate. Therefore we assume that an inventory holding cost c_i^{h} (> 0) is paid for all S_i spare parts of SKU i, which includes the spare parts in repair. So, the inventory holding costs for SKU i are equal to $c_i^{\mathrm{h}} S_i$. (Alternatively, one might assume a generalized inventory holding cost function consisting of a first part that is linear in the spare parts stock S_i and a second part that is linear in the average on-hand stock. This would only lead to a slightly modified analysis; see Problem 8.4.)

We pay a penalty cost c^{p} (> 0) per time unit per machine that is down because of a lack of spare parts. We assume that different backordered demands correspond to different machines. This is realistic for the same situations as for which the assumption of Poisson failure processes with constant rates is justified. Then the penalty costs may be measured via the total number of backordered parts. Let $EBO_i(S_i, \mathbf{U})$ denote the mean number of backordered demands of SKU i in steady state; this mean only depends on the steady-state number of failed parts of SKU i in the repair shop, which is determined by $\mathbf{U}$, and the basestock level S_i. The average penalty costs for systems that are down due to a temporarily lack of spare parts of SKU i are equal to $c^{\mathrm{p}} EBO_i(S_i, \mathbf{U})$. Hence, the average costs $C_i(S_i, \mathbf{U})$ for SKU i are given by

$$C_i(S_i, \mathbf{U}) = c_i^{\mathrm{h}} S_i + c^{\mathrm{p}} EBO_i(S_i, \mathbf{U}), \tag{8.3}$$

and the total average costs are given by $C(\mathbf{S}, \mathbf{U}) = \sum_{i \in I} C_i(S_i, \mathbf{U})$, where $\mathbf{S} = (S_1, \ldots, S_{|I|})$ is the vector consisting of all basestock levels.

Our objective is to choose the basestock policy $\mathbf{S}$ and the priority assignment $\mathbf{U}$ such that the total average costs are minimized. The mathematical formulation of this optimization problem is as follows:

$$\begin{aligned} (\mathrm{P}) \qquad \min \qquad & C(\mathbf{S}, \mathbf{U}) \\ \text{subject to } & \mathbf{S} \in \mathscr{S}, \\ & \mathbf{U} \in \mathscr{U}, \end{aligned}$$

where $\mathscr{S} = \{\mathbf{S} = (S_1, \ldots, S_{|I|}) \mid S_i \in \mathbb{N}_0, \forall i \in I\}$ denotes the solution space for the basestock policies $\mathbf{S}$ and $\mathscr{U} = \{\mathbf{U} = (u_{i,k})_{i \in I, k \in K} \mid u_{i,k} \in \{0, 1\}, \forall i \in I \text{ and } \forall k \in K;\ \sum_{k \in K} x_{i,k} = 1, \forall i \in I\}$ denotes the solution space for the priority assignments $\mathbf{U}$. Problem (P) is a non-linear integer optimization problem. The optimal costs of Problem (P) are denoted by C_{P}.

Because the penalty cost parameter c^{p} is the same for all SKU's, Problem (P) is closely related to the same problem with a constraint for the aggregate mean number of backorders instead of penalty costs. More precisely, the minimization of the Lagrangian function of the latter problem is equivalent to Problem (P); see Remark 8.1. This implies that working with SKU-independent penalty cost parameter leads to the same effects for optimizing the basestock policy and the priority assignment as working with a constraint for the aggregate mean number of backorders. In this chapter, we prefer to work with penalty costs because this avoids integrality effects when comparing the performance of priority assignments (plus corresponding optimal basestock policies) obtained by the various heuristics.

Remark 8.1. Suppose that one has a constraint for the aggregate mean number of backorders instead of penalty costs. Then the average costs for SKU i are given by $\hat{C}_i(S_i) = c_i^{\mathrm{h}} S_i$, $i \in I$, and the total average costs are equal to $\hat{C}(\mathbf{S}) = \sum_{i\in I} \hat{C}_i(S_i)$ (now the costs only depend on the basestock levels). The aggregate mean number of backorders is given by $EBO(\mathbf{S},\mathbf{U}) = \sum_{I\in I} EBO_i(S_i,\mathbf{U})$ and its target level is given by EBO^{obj}. We obtain the following optimization problem:

$$
\begin{aligned}
\text{(Q)} \qquad \min \qquad & \hat{C}(\mathbf{S}) \\
\text{subject to } & EBO(\mathbf{S},\mathbf{U}) \le EBO^{\mathrm{obj}}, \\
& \mathbf{S} \in \mathscr{S}, \\
& \mathbf{U} \in \mathscr{U},
\end{aligned}
$$

The Lagrangian function of Problem (Q) is defined as

$$L(\mathbf{S},\mathbf{U},\lambda) := \hat{C}(\mathbf{S}) + \lambda (EBO(\mathbf{S},\mathbf{U}) - EBO^{\mathrm{obj}}).$$

Minimizing $L(\mathbf{S},\mathbf{U},\lambda)$ for a given λ is equivalent to minimizing $\hat{C}(\mathbf{S}) + \lambda EBO(\mathbf{S},\mathbf{U})$. The latter problem is equivalent to Problem (P) with $c^{\mathrm{p}} = \lambda$. This shows the close relationship between Problem (P) and Problem (Q).

8.3 Evaluation

In this section, we describe the exact evaluation of the costs for a given basestock policy $\mathbf{S}$ and priority assignment $\mathbf{U}$. By (8.3), we can determine the average costs if we are able to determine the mean numbers of backordered parts $EBO_i(S_i,\mathbf{U})$ per SKU $i \in I$. To obtain those numbers, we first determine the steady-state distributions for the numbers of parts in repair per priority class and per SKU, respectively.

Let p_j^k be the probability that the number of parts in repair of priority class $k \in K$ equals $j \in \mathbb{N}_0$ at an arbitrary moment. The corresponding mean number of parts in repair is denoted by L^k. Because all repair times are identically distributed and our priority rule is preemptive, for the analysis of priority class k, all classes with a lower priority can be ignored, and class k itself may be ignored when higher priority classes are considered.

For priority class 1, there are no higher classes, and its behavior follows from the standard $M|M|1$ queueing system with arrival rate M_1 and service rate μ. The probabilities p_j^1, $j \in \mathbb{N}_0$, are equal to

$$p_j^1 = (1-\rho_1)\rho_1^j, \tag{8.4}$$

where $\rho_1 = M_1/\mu$ is the workload of priority class 1. Further, $L^1 = \rho_1/(1-\rho_1)$.

For each priority class $k > 1$, it holds that the server is working on higher priority classes during certain time intervals. The order in which the parts of these classes are repaired (or served) is irrelevant for class k, and we see all these parts of higher priority classes as one aggregated class with workload $\rho_k^{\mathrm{h}} = (\sum_{l=1}^{k-1} M_l)/\mu$. Further, in the time intervals that there are no parts present of the aggregated class, the server is completely available for class k. The workload of class k itself is $\rho_k = M_k/\mu$. The steady-state probabilities p_j^k now follow from the analysis of the two-queue preemptive priority model with the aggregated class and class k as the highest and lowest priority class, respectively. This system has been solved by Miller [8]; see also [13] for slightly simpler formulas, which we will use. These simpler formulas of [13] are obtained by the application of the matrix-analytic approach for $M|G|1$-type queueing systems (cf. Neuts [10]). The behavior of the two-queue preemptive priority model is described by a Markov process with states $(n^{\mathrm{h}}, n^{\mathrm{l}})$, where n^{h} and n^{l} represents the number of high-priority and low-priority parts, respectively. The state space is divided into levels, where level $j \geq 0$ is defined as the set of states $(n^{\mathrm{h}}, n^{\mathrm{l}})$ with $n^{\mathrm{h}} = j$. We then obtain a special upper triangular structure for the so-called fundamental matrix $\mathbf{G}$ of the matrix-geometric approach, where all components on the diagonal are equal to a factor g_0^k and all components on the j-th co-diagonal are equal to g_j^m, $j \geq 1$. The factors g_j^m represent first passage probabilities: g_j^k represents the probability that the number of low-priority jobs has increased by j at the first passage to level $n^{\mathrm{h}} - 1$ when starting in state $(n^{\mathrm{h}}, n^{\mathrm{l}})$; for further details, see [13]. Here, we simply need the formulas to compute the probabilities g_j^k:

$$g_0^k = \frac{(1+\rho_k^{\mathrm{h}}+\rho_k) - \sqrt{(1+\rho_k^{\mathrm{h}}+\rho_k)^2 - 4\rho_k^{\mathrm{h}}}}{2\rho_k^{\mathrm{h}}}, \tag{8.5}$$

$$g_{j+1}^k = \frac{\rho_k g_j^k + \rho_k^{\mathrm{h}} \sum_{l=1}^{j} g_l^k g_{j+1-l}^k}{(1+\rho_k^{\mathrm{h}}+\rho_k) - 2\rho_k^{\mathrm{h}} g_0^k}, \quad j \in \mathbb{N}_0, \tag{8.6}$$

where, by convention, $\sum_{l=1}^{j} g_l^k g_{j+1-l}^k = 0$ for $j = 0$. Next, the probabilities p_j^k are obtained by the following recursive formulas:

$$p_0^k = (1-\rho_k^{\mathrm{h}}-\rho_k) + \frac{\rho_k^{\mathrm{h}}}{\rho_k}(1-\rho_k^{\mathrm{h}}-\rho_k)(1-g_0^k), \tag{8.7}$$

$$\begin{aligned} p_j^k = \rho_k p_{j-1}^k &+ \rho_k^{\mathrm{h}} \sum_{l=0}^{j-1} \left[p_{j-1-l}^k \left(1 - \sum_{v=0}^{l} g_v^k \right) \right] \\ &+ \frac{\rho_k^{\mathrm{h}}}{\rho_k}(1-\rho_k^{\mathrm{h}}-\rho_k)\left(1 - \sum_{v=0}^{j} g_v^k\right), \quad j \in \mathbb{N}. \end{aligned} \tag{8.8}$$

For L^k, it holds that

$$L^k = \frac{\rho_k}{(1-\rho_k^{\mathrm{h}})(1-(\rho_k^{\mathrm{h}}+\rho_k))}.$$

In these formulas, both ρ_k^{h} and ρ_k are assumed to be strictly positive. If $\rho_k^{\mathrm{h}} = 0$, then there are no higher priority jobs and thus the steady-state probabilities and L^k are as in an $M|M|1$ system (cf. (8.4)). If $\rho_k^{\mathrm{h}} > 0$ and $\rho_k = 0$, then it simply holds that $p_0^k = 1$ and $p_j^k = 0$ for all $j > 0$ (which also follows from (8.7) and (8.8) when taking $\rho_k \downarrow 0$), and $L^k = 0$.

Next, we determine the distributions for the numbers of parts in the repair shop per SKU. Define $P_j^i(\mathbf{U})$ as the probability that j parts of SKU i are present in the repair shop, i.e., $P_j^i(\mathbf{U}) = P\{X_i = j\}$; we now explicitly denote that these probabilities depend on $\mathbf{U}$ because they are used in later sections where Problem (P) is analyzed. Recall that k_i denotes the class to which SKU i has been assigned. For priority class k_i, the total stream of arriving parts is a Poisson process that is constituted by Poisson arrival streams of all SKU's l that are assigned to class k_i. The intensity of the total stream is M_{k_i}. Because all parts arrive according to Poisson streams, each part of class k_i in the repair shop has a probability m_i/M_{k_i} that the part is of SKU i and the part is of one of the other SKU's in class k_i otherwise. Based on this property, we find that

$$P_j^i(\mathbf{U}) = \left[\sum_{l=j}^{\infty} \binom{l}{j} \left(\frac{m_i}{M_{k_i}} \right)^j \left(1 - \frac{m_i}{M_{k_i}} \right)^{l-j} p_l^{k_i} \right], \quad j \in \mathbb{N}_0. \tag{8.9}$$

Finally, for each SKU i, $BO_i = (X_i - S_i)^+$, and hence,

$$\begin{aligned} EBO_i(S_i, \mathbf{U}) &= \sum_{j=S_i+1}^{\infty} (j - S_i) P_j^i(\mathbf{U}) \\ &= \left(\frac{m_i}{M_{k_i}} \right) L^{k_i} - S_i + \sum_{j=0}^{S_i} (S_i - j) P_j^i(\mathbf{U}), \quad S_i \in \mathbb{N}_0. \end{aligned} \tag{8.10}$$

This completes the description of how to determine all $EBO_i(S_i, \mathbf{U})$ for a given basestock policy $\mathbf{S}$ and a given priority assignment $\mathbf{U}$. They are obtained via the (recursive) formulas (8.4)–(8.10).

Notice that in actual computations of the probabilities g_j^k, p_j^k, and $P_j^i(\mathbf{U})$, one has to apply truncation at a sufficiently high level. E.g., for the probabilities p_j^k, one computes them for $j = 0, 1, 2, \ldots$, and one stops as soon as the cumulative probability exceeds $1 - \varepsilon$, where ε is in the order of 10^{-6}–10^{-8}. The probabilities p_j^k for higher j are simply assumed to be zero. As a result, the infinite sum in formula (8.9) for $P_j^i(\mathbf{U})$ reduces to a finite sum, and then that formula can be applied directly for actual computations.

8.4 Optimization

In this section, we describe optimization methods for Problem (P). First, in Sect. 8.4.1, we reduce Problem (P) to a pure priority assignment problem by deriving how the basestock levels are optimized under a given priority assignment.

Next, in Sect. 8.4.2, we consider the optimization of the priority assignment. We describe an enumeration method to obtain an exact solution and we present four heuristic methods.

8.4.1 Optimization of the Basestock Policy

Suppose that a priority assignment $\mathbf{U}$ is given. Then Problem (P) becomes a multi-item optimization problem for the basestock policy $\mathbf{S}$ only:

$$\text{(P}(\mathbf{U}))\qquad \min \qquad C(\mathbf{S},\mathbf{U})$$

$$\text{subject to } \mathbf{S}\in\mathscr{S}.$$

This problem decomposes into single-item optimization problems ($\text{P}_i(\mathbf{U})$) per SKU i:

$$(\text{P}_i(\mathbf{U}))\qquad \min \qquad C_i(S_i,\mathbf{U})$$

$$\text{subject to } S_i\in\mathbb{N}_0.$$

Below we show that these problems can be solved along the same lines as a Newsvendor problem.

For the first order difference function of the objective function of Problem ($\text{P}_i\mathbf{U}$)), we find

$$\begin{aligned}\Delta_1 C_i(S_i,\mathbf{U})) &= C_i(S_i+1,\mathbf{U})) - C_i(S_i,\mathbf{U}))\\ &= c_i^{\text{h}} + c^{\text{p}}\left[-(S_i+1)+S_i+\sum_{j=0}^{S_i+1}(S_i+1-j)P_j^i(\mathbf{U})\right.\\ &\qquad \left. -\sum_{j=0}^{S_i}(S_i-j)P_j^i(\mathbf{U})\right]\\ &= -(c^{\text{p}}-c_i^{\text{h}}) + c^{\text{p}}\sum_{j=0}^{S_i}P_j^i(\mathbf{U}),\;\; S_i\in\mathbb{N}_0.\end{aligned}$$

This shows that $\Delta_1 C_i(S_i,\mathbf{U}))$ is increasing on its whole domain, and thus $C_i(S_i,\mathbf{U})$ is convex. Hence, $C_i(S_i,\mathbf{U})$ is minimized at the first point where $\Delta_1 C_i(S_i,\mathbf{U}))\geq 0$, i.e., at the smallest $S_i\in\mathbb{N}_0$ for which

$$\sum_{j=0}^{S_i}P_j^i(\mathbf{U})\geq\frac{c^{\text{p}}-c_i^{\text{h}}}{c^{\text{p}}}. \tag{8.11}$$

This optimal point is denoted by $S_i^*(\mathbf{U})$. The ratio $(c^{\text{p}}-c_i^{\text{h}})/c^{\text{p}}$ is the Newsvendor ratio for our problem; this ratio is different from the well-known Newsvendor ratio $c^{\text{p}}/(c^{\text{p}}+c_i^{\text{h}})$ because in our problem inventory holding costs are also paid for parts in repair. The sum $\sum_{j=0}^{S_i} P_j^i(\mathbf{U})$ is the fraction of time that there is no backlog for SKU i. Notice that $S_i^*(\mathbf{U}) = 0$ if $c_i^{\mathrm{h}} \geq c^{\mathrm{p}}$, and $S_i^*(\mathbf{U}) \rightarrow \infty$ if $c_i^{\mathrm{h}}/c^{\mathrm{p}} \downarrow 0$.

The basestock policy with basestock levels $S_i^*(\mathbf{U})$ is the optimal basestock policy for a given priority assignment $\mathbf{U}$ and is denoted by $\mathbf{S}^*(\mathbf{U})$. By substitution of this policy in Problem (P), Problem (P) reduces to the following optimization problem for the priority assignments $\mathbf{U}$:

$$\text{(P')} \qquad \min \qquad C(\mathbf{S}^*(\mathbf{U}), \mathbf{U})$$

$$\text{subject to } \mathbf{U} \in \mathscr{U}.$$

8.4.2 Optimization of the Priority Assignment

In this section, we describe solution methods for Problem (P'), i.e., for determining a (close-to-)optimal priority assignment. We first describe enumeration which leads to an exact solution. After that, we describe four heuristics.

Exact optimization of the assignment of the repair priorities is obtained by checking all possible priority assignments, i.e., by *enumeration*. Each evaluation of a given assignment requires that the probabilities $P_j^i(\mathbf{U})$, the optimal basestock levels $S_i^*(\mathbf{U})$, and expected numbers of backorders $EBO_i(S_i^*(\mathbf{U}), \mathbf{U})$ are determined via the formulas (8.9), (8.11), and (8.10). Since we have $|I|$ SKU's that are assigned to $|K|$ priority classes, the total solution space consists of $|K|^{|I|}$ different solutions (some solutions in the solution space are equivalent to others, but that is only a small fraction). This implies that the order of the computation time for enumeration is $|K|^{|I|}$, which is exponential as a function of $|I|$. Hence, enumeration will quickly result into too large computation times, and it makes sense to develop heuristics with polynomial computation times, so that also instances with many SKU's can be solved. Below, we develop these heuristics, and the enumeration algorithm will be used to study the optimality gap of these heuristics.

Many studies on optimal priority assignments in production systems (cf. Buzacott and Shanthikumar [2]) suggest that SKU's with higher costs should have a higher priority than the ones with lower costs. This seems also logical for our problem, as higher priorities for expensive SKU's are expected to lead to lower basestock levels and thus significantly lower inventory holding costs. At the same time, lower priorities for cheap SKU's will lead to higher basestock levels for them, but that will have only a limited effect on the inventory holding costs. This leads to the formulation of the first heuristic, called *Heuristic 1 (H1)*. From now on, w.l.o.g., we assume that the SKU's are ordered such that $c_1^{\mathrm{h}} \geq c_2^{\mathrm{h}} \geq \ldots \geq c_{|I|}^{\mathrm{h}}$. Next, we define an *ordered assignment* as an assignment under which $k_1 \leq k_2 \leq \ldots \leq k_{|I|}$; i.e., for each pair of SKU's i and i' with $i < i'$ it holds that SKU i is assigned to the same or a higher priority class than SKU i'. Heuristic 1 evaluates all ordered assignments and selects the best ordered assignment. Each ordered assignments is found by placing $|K| - 1$ gates in a row consisting of all SKU's $1, \ldots, |I|$; the SKU's left of the first gate belong to class 1, the SKU's between the first and second gate belong to class 2, and so on. Hence, the number of ordered assignments is equal to

$$\binom{|I|+|K|-1}{|K|-1} = \frac{(|I|+|K|-1)\cdot(|I|+|K|-2)\cdot\ldots\cdot(|I|+1)}{(|K|-1)!}.$$

As a result, for a given number of priority classes $|K|$, Heuristic 1 is polynomial in $|I|$ and the order of the computation time is $O(|I|^{|K|-1})$. Notice that the computation time is linear in $|I|$ in the special case of $|K|=2$ priority classes.

A faster variant of Heuristic 1 is *Heuristic 2 (H2)*, which is a local search algorithm within the class of all ordered assignments. Initially, we assign all SKU's to priority class 1. In each iteration step, we consider neighbors obtained by moving the SKU with the highest index in a non-empty class $k < |K|$ to class $k+1$. Hence, each iteration requires $\leq |K|-1$ evaluations of neighbors. The number of iterations is bounded from above by $(|K|-1)|I|$ as the number of times to move a given SKU to the next priority class is at most $|K|-1$. Therefore, the order of computation time for Heuristic 2 is $(|K|-1)^2|I|$. This order is linear in $|I|$ for any given $|K|$, while Heuristic 1 has only a linear computation time for $|K|=2$.

The following example shows that the optimal priority assignment also depends on the workloads of the SKU's, to some extent, and that therefore Heuristics 1 and 2 may lead to suboptimal solutions.

Example 8.1. Consider a situation with $|I|=2$ SKU's, $|K|=2$ priority classes, and

$$m_1 = 0.75,\ \ m_2 = 0.15,\ \ \mu = 1,$$
$$c_1^{\mathrm{h}} = 0.51,\ \ c_2^{\mathrm{h}} = 0.49,\ \ c^{\mathrm{p}} = 1.$$

In this case 4 different priority assignments exist, which we denote as $\mathbf{U}_1, \mathbf{U}_2, \mathbf{U}_3, \mathbf{U}_4$. For these assignments the optimal basestock levels and costs are given in Table 8.1. Notice that under $\mathbf{U}_3$ and $\mathbf{U}_4$, both SKU's are in the same priority class; these solutions are equivalent and they correspond to using the standard FCFS discipline in the repair shop.

The results in Table 8.1 show that priority assignment $\mathbf{U}_2$ is the unique optimal solution, but this solution is *not* an ordered assignment. Hence, Heuristics 1 and 2 would lead to suboptimal solutions in this case (Heuristic 1 would produce $\mathbf{U}_3$ or $\mathbf{U}_4$ as solution and Heuristic 2 would lead to $\mathbf{U}_3$). Apparently, in this example, it is optimal to place the SKU with the lowest workload (= lowest arrival rate) in the highest priority class instead of the most expensive SKU. As we know from the theory of priority queues that has little effect on the mean repair leadtime for the SKU with the high workload, while it has a strong reducing effect on the mean repair leadtime for the SKU with the low workload.

Based on the above insights we define improved versions of Heuristics 1 and 2. To both algorithms, we add a local search, in which we also allow non-ordered assignments. In this local search, we distinguish two types of neighbors:

Table 8.1 Computational results for Example 8.1

Priority assignment	$\mathbf{U}_1 = \begin{pmatrix} 1 & 0 \\ 0 & 1 \end{pmatrix}$	$\mathbf{U}_2 = \begin{pmatrix} 0 & 1 \\ 1 & 0 \end{pmatrix}$	$\mathbf{U}_3 = \begin{pmatrix} 1 & 0 \\ 1 & 0 \end{pmatrix}$	$\mathbf{U}_4 = \begin{pmatrix} 0 & 1 \\ 0 & 1 \end{pmatrix}$
$(S_1^*(\mathbf{U}_i), S_2^*(\mathbf{U}_i))$	(2,3)	(6,0)	(5,1)	(5,1)
$C(\mathbf{S}^*(\mathbf{U}_i), \mathbf{U}_i)$	8.22	7.91	7.95	7.95

- We allow that an SKU i is moved from its current priority class k_i to class $k_i - 1$ (only if $k_i > 1$) or to class $k_i + 1$ (only if $k_i < |I|$);
- We allow that two SKU's i and i' with priority classes k_i and $k_{i'} > k_i$ are swapped if k_i and $k_{i'}$ are so-called neighboring classes, i.e., if $k_{i'} = k_i + 1$, or if $k_{i'} > k_i + 1$ and all priority classes k, $k_i < k < k_{i'}$, are empty.

Heuristic 3 (H3) consists of Heuristic 1 followed by this local search procedure. Similarly, *Heuristic 4 (H4)* consists of Heuristic 2 followed by this local search procedure. In this local search procedure the number of neighbors can be shown to be bounded by $2|I| + \frac{1}{2}|I|^2$ ($2|I|$ for the first type of neighbors and $\frac{1}{2}|I|^2$ for the second type of neighbors). However, there is no tight bound for the number of iterations. Hence, for the orders of computation time for the Heuristics 3 and 4, we only know that they are equal to or larger than the orders of computation time of the Heuristics 1 and 2.

When comparing the Heuristics 1–4 to each other with respect to accuracy, we know that Heuristic 1 dominates Heuristic 2. I.e., in all instances, the solution generated by Heuristic 1 will be at least equally good as the solution generated by Heuristic 2. Further, Heuristic 3 dominates Heuristics 1 and 2, and Heuristic 4 dominates Heuristic 2.

8.5 Computational Results

In this section, we define a large test bed in Sect. 8.5.1. Next, in Sect. 8.5.2, we present results for the optimality gap of the Heuristics 1–4 for instances with 15 SKU's and 2 priority classes. In Sect. 8.5.3, we compare the performance of the Heuristics 1–4 for larger instances with up to 50 SKU's and 5 priority classes. Finally, in Sect. 8.5.4, we investigate the costs savings that are obtained by using static priorities instead of the FCFS service discipline.

8.5.1 Test Bed

We use a factorial design for our test bed. The number of SKU's is chosen equal to 15, 25, and 50, and we have 2, 3, 4, and 5 priority classes. Next, regarding the holding cost, we fix the highest holding cost ($c^{\mathrm{h}}_{\mathrm{max}}$) at 1,000. The lowest holding cost ($c^{\mathrm{h}}_{\mathrm{min}}$) is taken equal to 1, 10, and 100. Hence the ratio of highest and lowest inventory holding cost varies from 10 to 1,000, which are common ratios for spare parts. The dependence between the inventory holding cost parameters and demand rates for the SKU's is another factor taken into account. We test three different variants for this dependence. In all three variants, we choose the demand rates m_i for all SKU's as independent samples from a uniform distribution on [1,100]. Then, in the first variant, the holding cost parameters and demand rates are chosen independently. In this case, the holding cost parameters are picked as independent samples from a uniform distribution on $[c^{\mathrm{h}}_{\min}, c^{\mathrm{h}}_{\max}]$ (see Fig. 8.1a). In the second variant, we assume a hyperbolic relation between demand rates and holding cost parameters (see Fig. 8.1b), as this reflects what one typically sees in practice. The holding cost parameters m_i are obtained by the following function:

$$c^{\mathrm{h}}_i = \max\left\{c^{\mathrm{h}}_{\min}, a\frac{110}{10+m_i} + b + \xi_i\right\}. \tag{8.12}$$

In this function, $a = \frac{1}{9}(c^{\mathrm{h}}_{\max} - c^{\mathrm{h}}_{\min})$ and $b = \frac{10}{9}c^{\mathrm{h}}_{\min} - \frac{1}{9}c^{\mathrm{h}}_{\max}$, and $\xi_i \in U[-v, v]$ with $v = \frac{1}{40}(c^{\mathrm{h}}_{\max} - c^{\mathrm{h}}_{\min})$. Notice that $c^{\mathrm{h}}_i = c^{\mathrm{h}}_{\min}$ if $m_i = 100$ and $\xi_i = 0$, and $c^{\mathrm{h}}_i = c^{\mathrm{h}}_{\max}$ if $m_i = 1$ and $\xi_i = 0$. In the third variant,we assume the same relation as in the second variant, but we include some SKU's that are picked differently (see Fig. 8.1c). For most of the SKU's ($\frac{2}{3}|I|$), we assume the hyperbolic dependence as in the second variant (cf. (8.12)), for some of the SKU's ($\frac{2}{9}|I|$) we take low demand rates and low inventory holding cost parameters, and for some SKU's ($\frac{1}{9}|I|$) we take high demand rates and high inventory holding cost parameters (see Fig. 8.1c). Further, the mean repair time is chosen such that the utilization rate of the repair shop, defined by $\rho := (\sum_{i\in I} m_i)/\mu$, is equal to 0.7, 0.82, 0.9, and 0.95, respectively. Finally, for the penalty cost parameter c^{p}, we choose the values 1,000, 10,000, and 100,000. In Table 8.2, we summarize our choices for all parameters.

The total number of all possible combinations for these parameters is $3 \times 4 \times 3 \times 1 \times 3 \times 4 \times 3 = 1{,}296$. In addition, for each of the three variants for the m_i and c^{h}_i we generate 5 sets of values as there are uniform distributions involved in the generation of these values (notice that optimal solutions may be very sensitive for small changes in input parameters as we are dealing with an integer optimization problem). This gives us in total $1{,}296 \times 5 = 6{,}480$ instances.

8.5.2 Optimality Gap of the Heuristics

The quality of the Heuristics 1–4 may be tested by comparison of the costs of the generated solutions to the optimal costs. For sufficiently small values of $|K|$ and $|I|$, the optimal costs may be obtained by enumeration. We applied the enumeration method and all 4 heuristics for all instances with $|I| = 15$ SKU's and $|K| = 2$ priority classes. Per instance and heuristic, we computed the optimality gap G_k as the relative distance of the costs of the heuristic solution to the optimal costs:

$$G_k = \frac{C_k - C_{\mathrm{P}}}{C_{\mathrm{P}}}, \quad k \in \{1,2,3,4\},$$

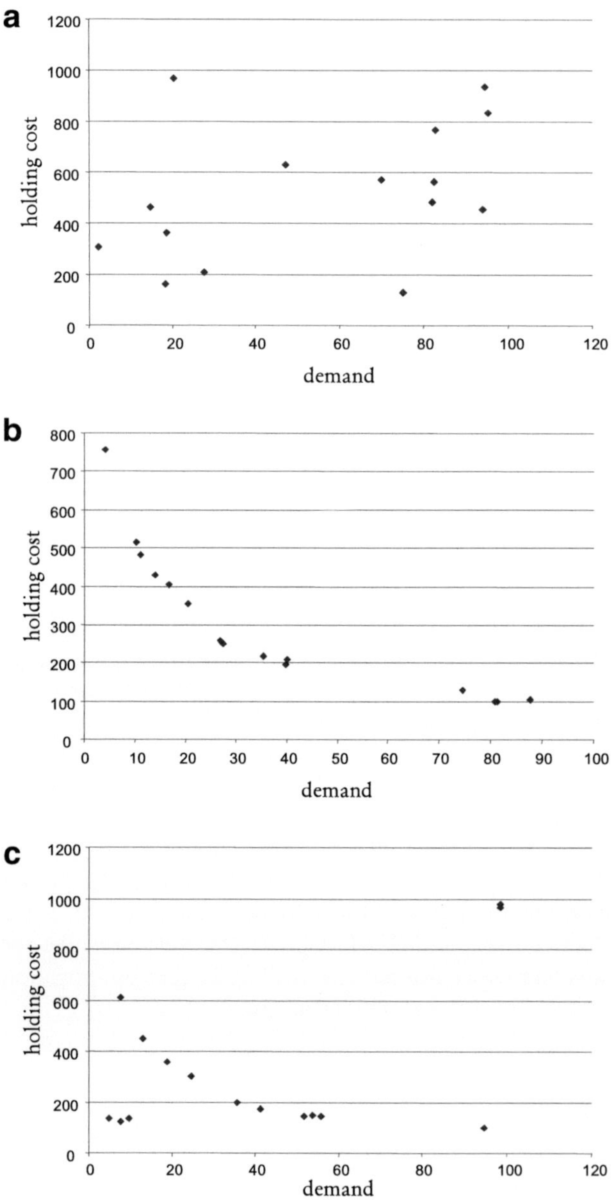

Fig. 8.1 Relations between inventory holding cost parameters and demand rates

Table 8.2 Overview of parameter choices for the test bed

Name of parameter	No. val.	Values
No. of SKU's ($\|I\|$)	3	15, 25, 50
No. of priority class. ($\|K\|$)	4	2, 3, 4, 5
Lowest hold. cost ($c^{\mathrm{h}}_{\min}$)	3	1, 10, 100
Highest hold. cost ($c^{\mathrm{h}}_{\max}$)	1	1,000
Demand rates (m_i) Holding costs (c_i^{h})	3	1) $m_i \in U[1,100]$, $c_i^{\mathrm{h}} \in U[c^{\mathrm{h}}_{\min}, c^{\mathrm{h}}_{\max}]$, 2) $m_i \in U[1,100]$, $m_i = c^{\mathrm{h}}(m_i)$ (see (8.12)), 3) $\begin{cases} m_i \in U[1,100], & c_i^{\mathrm{h}} = c^{\mathrm{h}}(m_i), & 1 \le i \le \frac{2}{3}\|I\| \\ m_i \in U[1,10], & c_i^{\mathrm{h}} \in U[c^{\mathrm{h}}_{\min}, c^{\mathrm{h}}_{\min}+v], & \frac{2}{3}\|I\| < i \le \frac{8}{9}\|I\| \\ m_i \in U[90,100], & c_i^{\mathrm{h}} \in U[c^{\mathrm{h}}_{\max}-v, c^{\mathrm{h}}_{\max}], & \frac{8}{9}\|I\| < i \le \|I\| \end{cases}$
Utilization rate (ρ)	4	0.7, 0.82, 0.9, 0.95
Penalty cost (c^{P})	3	1,000, 10,000, 100,000

Table 8.3 Average optimality gaps G_k (for the test bed limited to $|I| = 15$ and $|K| = 2$)

(a) Aver. G_k as a function of $c^{\mathrm{h}}_{\min}$

	1 (%)	10 (%)	100 (%)
H1	5.1	5.4	5.9
H2	23.3	24.7	19.7
H3	1.1	1.3	0.9
H4	9.7	9.4	7.9

(b) Aver. G_k as a function of m_i/c_i^{h}

	Case 1 (%)	Case 2 (%)	Case 3 (%)
H1	5.6	5.0	5.8
H2	13.8	26.7	27.2
H3	1.0	1.5	0.8
H4	5.0	11.6	10.4

(c) Aver. G_k as a function of ρ

	0.7 (%)	0.82 (%)	0.9 (%)	0.95 (%)
H1	10.8	5.7	3.6	1.8
H2	35.8	27.8	18.8	7.9
H3	1.8	1.0	1.1	0.5
H4	9.5	13.6	9.8	3.1

(d) Aver. G_k as a function of c^{P}

	1,000 (%)	10,000 (%)	100,000 (%)
H1	0.8	4.7	10.9
H2	5.9	27.8	34.0
H3	0.0	0.3	3.0
H4	1.4	10.5	15.1

(e) Aver. G_k over all instances

H1	5.5 %
H2	22.6 %
H3	1.1 %
H4	9.0 %

where C_k denotes the costs of the solution generated by Heuristic $k \in \{1,2,3,4\}$. The averages over different subsets of instances for these optimality gaps are listed in Table 8.3. E.g., in part (a) of this table, the average optimality gaps for all instances with $c^{\mathrm{h}}_{\min} = 1$, 10, and 100 are given in the second, third, and fourth column, respectively. This shows how the optimality gap depends on the choice for $c^{\mathrm{h}}_{\min}$. In the last part of the table, part (e), the averages over all instances are given.

The results in part (e) show that Heuristic 3 has the best performance. This heuristic performs very well with an average optimality gap of 1.1 %. Heuristics 1 and 4 perform reasonably well, and the performance of Heuristic 2 is relatively bad. When we look into further detail in the parts (a)–(d), we observe that the gap for Heuristic 3 is not so sensitive for the choice of the lowest inventory holding cost

$c^{\mathrm{h}}_{\mathrm{min}}$. It is somewhat sensitive for the relation that is assumed between the demand rates and inventory holding cost parameters. Further, the gap decreases significantly for increasing workloads ρ, and increases significantly for increasing values of the penalty cost parameter c^{p}. For the Heuristics 1, 2, and 4, we see a similar behavior in the parts (a), (c), and (d), while the picture in part (b) is more mixed.

The average computation times per instance for the Heuristics 1–4, on a PC with a PIII-1,000 MHz processor, were equal to 0.15, 0.07, 1.47, and 1.43 s, respectively. So, these times were small for all instances.

8.5.3 Comparison of the Heuristics for Large Instances

In this section, we compare the performance of the Heuristics 1–4 for the full test bed. This time we cannot compare to enumeration, since enumeration requires too much computation time for the instances with large $|I|$ or $|K|$. Because in the previous section Heuristic 3 was the best algorithm, here the performance of the other three heuristics is expressed relative to Heuristic 3. Per instance, we compute the relative distance of the costs of the heuristic solution generated by Heuristic 1 to the costs of the heuristic solution generated by Heuristic 3; and, similarly for the Heuristics 2 and 4. By definition, nonnegative relative differences are obtained for the Heuristics 1 and 2 as these algorithms are dominated by Heuristic 3. In the case of Heuristic 4, also negative relative differences may be obtained. The results are listed in Table 8.4, where again averages over different subsets of instances are listed.

We see that the performance of Heuristic 1 is on average only 5.1 % worse relative to Heuristic 3, Heuristic 4 is 25.9 % worse, and Heuristic 2 is 46.0 % worse (see part (g)). Although Heuristic 4 is much worse than Heuristic 3 on average, Heuristic 4 generated a better solution than Heuristic 3 in 8 % of all instances. The performance of Heuristic 1 is rather stable when input parameters are varied. The performance of the Heuristics 2 and 4 strongly depends on various parameters, among which the number of SKU's.

Table 8.5 shows the computation times (in seconds) of the Heuristics 1–4 executed on a PC with a PIII-1,000 MHz processor. It shows that Heuristics 1 and 3 have limited computation times for instances with $|K| = 2$ or 3 priority classes, and the computation times become much larger for $|K| = 4$ and 5. Heuristics 2 and 4 have small computation times for all M.

From the above results for our full test bed, we conclude that Heuristic 3 is most appropriate, at least for instances with 2 or 3 priority classes. Heuristic 4 may constitute an attractive alternative if computation times for Heuristic 3 become too large.

Table 8.4 Relative difference between total costs obtained by the four heuristics

(a) Rel. diff. as a function of $c^{\mathrm{h}}_{\mathrm{min}}$

	1 (%)	10 (%)	100 (%)
H1 vs. H3	5.4	5.8	4.2
H2 vs. H3	54.6	50.5	33.0
H4 vs. H3	30.7	28.2	18.8

(b) Rel. diff. as a function of m_i/c^{h}_i

	Case 1 (%)	Case 2 (%)	Case 3 (%)
H1 vs. H3	5.7	2.4	7.3
H2 vs. H3	31.5	57.7	48.9
H4 vs. H3	16.0	38.7	22.9

(c) Rel. diff. as a function of ρ

	0.7 (%)	0.82 (%)	0.9 (%)	0.95 (%)
H1 vs. H3	6.3	5.6	4.5	4.1
H2 vs. H3	43.5	48.1	46.5	46.0
H4 vs. H3	16.8	26.6	29.4	30.7

(d) Rel. diff. as a function of c^{p}

	1,000 (%)	10,000 (%)	100,000 (%)
H1 vs. H3	0.9	3.2	11.2
H2 vs. H3	12.0	63.9	62.2
H4 vs. H3	4.0	35.9	37.7

(e) Rel. diff. as a function of $|K|$

	2 (%)	3 (%)	4 (%)	5 (%)
H1 vs. H3	6.0	5.8	5.4	5.1
H2 vs. H3	35.9	45.6	46.2	46.0
H4 vs. H3	18.4	25.5	26.0	25.9

(f) Rel. diff. as a function of $|I|$

	15 (%)	25 (%)	50 (%)
H1 vs. H3	2.7	6.2	6.5
H2 vs. H3	33.7	44.8	59.6
H4 vs. H3	18.0	25.3	34.4

(g) Aver. rel. diff. over all instances

H1 vs. H3	5.1 %
H2 vs. H3	46.0 %
H4 vs. H3	25.9 %

Table 8.5 Average computation times (in seconds) of the Heuristics 1–4

(a) Heuristic 1

	$\|I\| = 15$	$\|I\| = 25$	$\|I\| = 50$
$\|K\| = 2$	0.15	0.28	0.69
$\|K\| = 3$	1.17	3.83	18.34
$\|K\| = 4$	6.75	37.74	367.70
$\|K\| = 5$	26.59	268.10	5,520.94

(b) Heuristic 2

	$\|I\| = 15$	$\|I\| = 25$	$\|I\| = 50$
$\|K\| = 2$	0.07	0.10	0.17
$\|K\| = 3$	0.12	0.18	0.30
$\|K\| = 4$	0.17	0.25	0.42
$\|K\| = 5$	0.22	0.32	0.54

(c) Heuristic 3

	$\|I\| = 15$	$\|I\| = 25$	$\|I\| = 50$
$\|K\| = 2$	1.47	4.75	29.25
$\|K\| = 3$	3.12	10.97	66.72
$\|K\| = 4$	9.13	46.47	429.54
$\|K\| = 5$	29.36	278.73	5,360.88

(d) Heuristic 4

	$\|I\| = 15$	$\|I\| = 25$	$\|I\| = 50$
$\|K\| = 2$	1.43	3.77	19.19
$\|K\| = 3$	2.10	5.57	25.99
$\|K\| = 4$	2.26	5.72	26.51
$\|K\| = 5$	2.31	5.79	26.73

8.5.4 Costs Savings Relative to FCFS

In this section, we investigate the costs savings that are obtained via the use of priority classes. We take again the full test bed, and for all instances, we apply our best heuristic, Heuristic 3 (recall that our experiment in Sect. 8.5.2 showed a small optimality gap of 1.1 %), and we determine the optimal costs that are obtained if the FCFS service discipline is used in the repair shop. The latter case corresponds to using only one priority class, which is a special case of our model; notice that then only the basestock levels have to be optimized (cf. Sect. 8.4.1). By definition, the costs obtained via Heuristic 3 are lower than or equal to the optimal costs under

Table 8.6 Relative savings in total costs via static priorities

(a) Rel. savings as a function of $c^{\rm h}_{\rm min}$

	1 (%)	10 (%)	100 (%)
$\lvert K\rvert = 2$	49.2	46.0	33.1
$\lvert K\rvert = 3$	54.2	49.8	34.8
$\lvert K\rvert = 4$	54.5	50.1	34.8
$\lvert K\rvert = 5$	54.5	50.1	34.8

(b) Rel. savings as a function of $m_i/c^{\rm h}_i$

	Case 1 (%)	Case 2 (%)	Case 3 (%)
$\lvert K\rvert = 2$	32.2	45.9	50.2
$\lvert K\rvert = 3$	35.9	50.1	52.9
$\lvert K\rvert = 4$	36.2	50.3	52.9
$\lvert K\rvert = 5$	36.2	50.4	52.9

(c) Rel. savings as a function of ρ

	0.7 (%)	0.82 (%)	0.9 (%)	0.95 (%)
$\lvert K\rvert = 2$	39.0	41.4	44.1	46.5
$\lvert K\rvert = 3$	40.2	44.1	48.0	52.9
$\lvert K\rvert = 4$	40.2	44.1	48.4	53.2
$\lvert K\rvert = 5$	40.2	44.1	48.4	53.2

(d) Rel. savings as a function of $c^{\rm p}$

	1,000 (%)	10,000 (%)	100,000 (%)
$\lvert K\rvert = 2$	39.9	49.1	39.2
$\lvert K\rvert = 3$	41.6	52.9	44.4
$\lvert K\rvert = 4$	41.7	53.1	44.6
$\lvert K\rvert = 5$	41.7	53.1	44.6

(e) Rel. savings as a function of $\lvert I\rvert$

	15 (%)	25 (%)	50 (%)
$\lvert K\rvert = 2$	38.9	42.4	46.9
$\lvert K\rvert = 3$	43.2	46.2	49.5
$\lvert K\rvert = 4$	43.6	46.4	49.4
$\lvert K\rvert = 5$	43.6	46.4	49.4

(f) Rel. savings over all instances

$\lvert K\rvert = 2$	42.8 %
$\lvert K\rvert = 3$	46.2 %
$\lvert K\rvert = 4$	46.5 %
$\lvert K\rvert = 5$	46.5 %

the FCFS discipline, and we study the relative savings in costs that are obtained via the solution of Heuristic 3. These numbers denote how profitable it may be to work with static priority classes. The results are listed in Table 8.6.

From Table 8.6, we draw the following conclusions:

- First of all, from part (f), we conclude that the average relative savings are more than 40 %. In 42 % of the instances, the savings were between 40 % and 60 %, and in 24 % they were even more than 60 %. These percentages imply that static priorities may lead to enormous costs savings in absolute terms for real-life situations (see also Sect. 8.6).
- Second, the results in part (f) show that, for an average instance, the maximal savings are around 46.5 %, and these savings are reached at 4 priority classes. More than 90 % of the maximal savings is already obtained when one works with 2 priority classes (42.6 % for $\lvert K\rvert = 2$ vs. 46.5 % as maximum). And, via 3 priority classes almost 100 % of the maximal savings is obtained. This is an important observation from both an algorithmic and practical point of view. It implies that effective and efficient heuristics are only needed for systems with 2 or 3 priority classes. Heuristic 3 meets this requirement. Further, in practice, implementing priority control rules with 2 or 3 priority classes will be much easier than rules with 4 or more priority classes.
- Third, the parts (a)–(e) show how the relative savings depend on the various input parameters. They strongly increase for increasing cost differences in inventory holding cost parameters (part (a)) and for increasing workloads (part (c)). Further, correlated demand rates and inventory holding cost parameters, as common in practice, lead to significantly larger savings than when these parameters are not correlated; see Cases 2 and 3 vs. Case 1 in part (b) of the table. Part (d) shows

that savings first increase and then decrease for increasing penalty costs. From part (e), we learn that the savings slightly increase as a function of the number of SKU's.

8.6 Stylized Cases

Based on the insights of the previous section, we executed an additional experiment with real life data of the Royal Netherlands Navy. This experiment was executed in 2004; see Dirkzwager [4]. We obtained data of so-called capacity groups that are responsible for preventive and corrective maintenance and for the repair of spare parts for different subsystems of technical systems installed at frigates. Reliable data that we could collect from their information systems were for the numbers of SKU's that were handled per capacity group, the prices of these SKU's (which we could easily translate to inventory holding cost parameters), and the total demand rates. No reliable data were available on (mean) repair times for SKU's. We therefore constructed one stylized case per capacity group. We pretended that at each capacity group there is a separate group of engineers that only works on repairs of spare parts, and further we make the same assumptions as in our model. Per case, we chose the $|I|$, m_i, and c_i^{h} according to the above data. The mean repair time was chosen such that the workload was equal to 0.8 and 0.9, respectively. We considered both the use of FCFS and the use of static priorities with 2 priority classes for the repair shop. We limited ourselves to 2 priority groups as that brings most of the maximal possible savings (cf. Sect. 8.5.4). For the system with static priorities, we applied Heuristic 3. Per case, the penalty cost parameter was tuned such that under the FCFS service discipline an aggregate fill rate β^{FCFS} of 90 % was obtained. It holds that

$$\beta^{\mathrm{FCFS}} = \sum_{i \in I} \left[\frac{m_i}{\sum_{l \in I} m_l} \sum_{j=0}^{S_i - 1} P_j^i(S_i, \mathbf{U}_{\mathrm{FCFS}}) \right],$$

where $\mathbf{U}_{\mathrm{FCFS}}$ corresponds to a priority assignment in which all SKU's are assigned to the same priority class. The larger the c^{p} is chosen, the larger the optimal basestock levels will be (cf. (8.11)), and thus the larger β^{FCFS}. We chose c^{p} as the lowest value for which $\beta^{\mathrm{FCFS}} \geq 0.90$.

In Table 8.7, the number of SKU's per case is given. In Fig. 8.2, the data of the inventory holding cost parameters and demand rates are plotted (for the lists with all values, see [4]). For the Cases 1–3, we see similar patterns as in the cases 2 and 3 of the test bed of Sect. 8.5. For Case 4, we see a rare pattern, with several SKU's that have high inventory holding costs and high demand rates, and there are no SKU's with low inventory holding costs and high demand rates. Also, in Cases 1–3, we see a large ratio between the highest and lowest inventory holding cost parameter (the ratio is 1,138 for Case 1, 457 for Case 2, and 458 for Case 3). In Case 4, this ratio is small (equal to 5).

Table 8.7 Overview of the results for the Royal Netherlands Navy for $|K| = 2$

	$\|I\|$	Workload	Penalty (c^{p}) (Euro/year)	C_{FCFS} (Euro/year)	C_{Prior} (Euro/year)	Relative reduction (%)
Case 1	67	0.8	3,596	3,239	1,723	47
		0.9	2,406	4,507	2,007	55
Case 2	99	0.8	13,958	25,701	15,612	39
		0.9	23,427	32,052	16,612	48
Case 3	29	0.8	124,624	124,694	69,975	44
		0.9	66,657	158,011	70,619	55
Case 4	13	0.8	136,249	196,215	176,989	10
		0.9	106,656	333,435	271,821	18

The main results for these cases are also presented in Table 8.7. In the fourth column, the tuned penalty cost parameter c^{p} is given. In the fifth and sixth column, we have listed the costs under FCFS and the use of 2 priority classes, respectively (these costs include penalty costs). The costs reduction obtained via priority classes is given in the last column.

We find that large costs reductions are obtained for Cases 1–3, with percentages varying from 39 % to 55 %. For workloads of 90 %, the savings are significantly larger than for workloads of 80 %. For Case 4, the costs reductions are relatively small. This is due to the small ratio between the highest and lowest inventory holding cost parameter in that case.

8.7 Concluding Remarks

The analysis and insights of this chapter stem from Adan et al. [1]. It has been shown, in a setting with a single stockpoint, multiple items, and a single-server repair shop, that in comparison to a FCFS control large costs reductions can be realized by using static priorities in the repair shop. Earlier, Hausman and Scudder [5] and Pyke [12] showed for the same setting that large reductions can be obtained via dynamic priorities (they used simulation and limited themselves to instances with low numbers of SKU's). Via dynamic priorities even larger costs reductions are possible. Tiemessen and Van Houtum [15] compared multiple dynamic priority policies, among which the policy of Hausman and Scudder [5] and a myopic allocation rule of Peña Perez and Zipkin [11], which was developed for a production system making multiple products to stock. Their comparison was under given basestock policies and for both systems with the same repair rate for all SKU's and systems with SKU-dependent repair rates. The myopic allocation rule performed the best in this study. Optimization of the basestock vectors seems well possible, also for systems with large numbers of SKU's, via a simulation-based procedure; see the procedure described by Tiemessen et al. [16] for a make-to-stock production system.

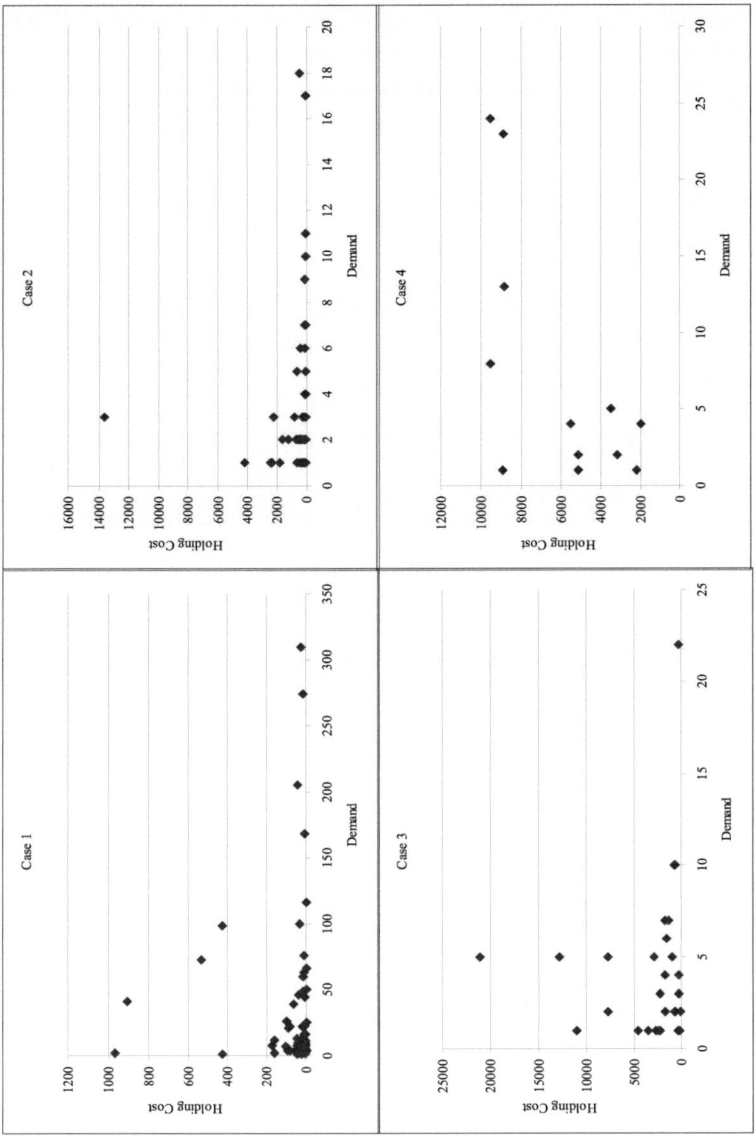

Fig. 8.2 Inventory holding cost parameters (in Euros/year) and yearly demand rates for Cases 1–4

The reducing effect of static priorities on total costs has also been studied for two-echelon, two-indenture systems with multi-server repair shops; see Sleptchenko et al. [14]. Further, Caggiano et al. [3] (see also [9]) developed an integrated model for real-time repair and inventory allocation decisions in a two-echelon repairable inventory system with one central repair facility. This is a finite-horizon, periodic-review, mathematical programming model that can be used in a rolling-horizon fashion at an operational decision level. A similar problem as the problem of this chapter, but with a fixed penalty cost per backordered demand (i.e., independent of the amount of time that a demand is backordered) has been studied by Kat and Avşar [7].

Problems

8.1. Let X be random variable that has a geometric distribution with success probability p, $0 \leq p < 1$ (see also Appendix "Fitting Discrete Distributions on the First Two Moments"). Let Y be a random variable with

$$P\{Y = y | X = x\} = \binom{x}{y} f^y (1-f)^{x-y} P\{X = x\}, \quad x, y \in \mathbb{N}_0, y \geq x,$$

where $f \in [0,1]$. Show that Y has a geometric distribution with success probability

$$p' = \frac{pf}{1-p+pf}.$$

Hint: You may use the following property. For $k \in \mathbb{N}_0$ and $0 \leq z < 1$, it holds that

$$\sum_{n=0}^{\infty} \binom{n+k}{k} z^n = \frac{1}{(1-z)^{k+1}}$$

(the sum on the lefthand side is the Taylor series of the function on the righthand side of this equation).

8.2. Consider Example 8.1 in Sect. 8.4. In this problem, we ask you to verify the listed results for priority assignment $\mathbf{U}_3$. Under this assignment, both SKU's are in priority class 1.

(a) Determine the probabilities p_j^1, $j \in \mathbb{N}_0$, and L^1.
(b) For SKU 1, determine the probabilities $P_j^1(\mathbf{U}_3)$, $j \in \mathbb{N}_0$, and $S_1^*(\mathbf{U}_3)$. Hint: Use the property given in Problem 8.1.
(c) For SKU 2, determine the probabilities $P_j^2(\mathbf{U}_3)$, $j \in \mathbb{N}_0$, and $S_2^*(\mathbf{U}_3)$.
(d) Determine $C_i(S_i^*(\mathbf{U}_3), \mathbf{U}_3)$, $i = 1, 2$, and $C(\mathbf{S}^*(\mathbf{U}_3), \mathbf{U}_3)$.

8.3. Consider the same model as in this chapter. However, we now assume that an emergency procedure is applied in case of a stockout. Each time that such a procedure is applied, one has a corresponding cost, which includes the costs for the downtime of a machine. In this problem, we consider a specific problem instance with $|I| = 2$ SKU's, $m_1 = m_2 = \frac{1}{3}$, $\mu = 1$, $S_1 = 1$, and $S_2 = 2$. At the repair shop, SKU 1 gets priority over SKU 2 (i.e., SKU 1 is in priority class 1 and SKU 2 is in priority class 2). We will determine how often the emergency procedure is applied.

(a) The behavior at the repair shop can be described by a Markov process with states (n_1, n_2), where n_i denotes the number of failed parts of SKU $i \in I$ in the repair shop. Notice that $0 \leq n_i \leq S_i$, $i \in I$. Depict the states, transitions, and transition rates of this Markov process.

(b) Determine the steady-state probabilities $\pi(n_1, n_2)$. Hint: When solving the steady-state equations, one can exploit the specific structure of the transitions for this Markov process. One can first express $\pi(1,0)$ as a factor times $\pi(0,0)$ by looking at the steady-state equation for state $(1,0)$. Next, $\pi(0,1)$ can be expressed as a factor times $\pi(0,0)$ via the steady-state equation for state (0,0); and so on.

(c) How often is the emergency procedure applied (per unit of time)?

8.4. In the model of this chapter, we have assumed that an inventory holding cost c_i^{h} is paid for all S_i spare parts of SKU i, $i \in I$. In this way, we have modeled the costs for the invested capital in spare parts. Let us now consider a more general situation with a second inventory holding cost parameter $\hat{c}_i^{\mathrm{h}}$. A cost $\hat{c}_i^{\mathrm{h}}$ is paid for on-hand stock of SKU i, $i \in I$. This second parameter can be used to model storage costs or other costs that only depend on the on-hand stock. This gives a model variant with a generalized costs structure.

(a) Give the formula for the average costs $C_i(S_i, \mathbf{U})$ under the generalized costs structure.

(b) It is easily seen that the evaluation of Sect. 8.3 remains applicable without any changes. In the optimization as described in Sect. 8.4, we do get a change. Derive a generalized characterization for the optimal basestock levels $S_i^*(\mathbf{U})$ under a given priority assignment $\mathbf{U}$.

(c) Is there anything else that needs to be changed within the methods described in Sect. 8.4?

References

1. Adan, I.J.B.F., Sleptchenko, A., Van Houtum, G.J.: Reducing costs of spare parts supply systems via static priorities. Asia-Pac. J. Oper. Res. **26**, 559–585 (2009). Copyright © 2009 by World Scientific Publishing Co. & Operational Research Society of Singapore
2. Buzacott, J.A., Shanthikumar, J.G.: Stochastic Models of Manufacturing Systems. Prentice Hall, Englewood Cliffs (1993)
3. Caggiano, K.E., Muckstadt, J.A., Rappold, J.A.: Integrated real-time capacity and inventory allocation for reparable service parts in a two-echelon supply system. Manuf. Serv. Oper. Manag. **8**, 292–319 (2006)
4. Dirkzwager, A.A.: On the use of priority classes in a capacitated model of repairable spare parts at the Royal Netherlands Navy. M.Sc. thesis, master program Industrial Engineering, Eindhoven University of Technology (2004). http://w3.tue.nl/en/services/library/digilib/publications_from_tue/masters_theses/
5. Hausman, W., Scudder, G.: Priority scheduling rules for repairable inventory systems. Manag. Sci. **28**, 1215–1232 (1982)
6. Hopp, W.J., Spearman, M.L.: Factory Physics, 3rd edn. Irwin McGraw-Hill, Boston (2008)

7. Kat, B., Avşar, Z.M.: Using aggregate fill rate for dynamic scheduling of multi-class systems. Ann. Oper. Res. **182**, 87–117 (2011)
8. Miller, D.R.: Computation of steady-state probabilities for $M/M/1$ priority queues. Oper. Res. **29**, 945–958 (1981)
9. Muckstadt, J.A.: Analysis and Algorithms for Service Parts Supply Chains. Springer, New York (2005)
10. Neuts, M.F.: Matrix-Geometric Solutions in Stochastic Models; An Algorithmic Approach. Dekker, New York (1989)
11. Peña Perez, A., Zipkin, P.: Dynamic scheduling rules for a multiproduct make-to-stock queue. Oper. Res. **45**, 919–930 (1997)
12. Pyke, D.F.: Priority repair and dispatch policies for repairable-item logistics systems. Nav. Res. Logist. **37**, 1–30 (1990)
13. Sleptchenko, A., Adan, I.J.B.F., Van Houtum, G.J.: Joint queue length distribution of multi-class, single-server queues with preemptive priorities. EURANDOM report 2004-045, Eindhoven University of Technology (2004)
14. Sleptchenko, A., Van Harten, A., Van der Heijden, M.C.: Using repair priorities to reduce stock investment in spare part networks. Eur. J. Oper. Res. **163**, 733–750 (2005)
15. Tiemessen, H.G.H., Van Houtum, G.J.: Reducing costs of repairable inventory supply systems via dynamic scheduling. Int. J. Prod. Econ. **143**, 478–488 (2013)
16. Tiemessen, H.G.H., Fleischmann, M., Van Houtum, G.J.: Dynamic control in multi-item production/inventory systems. Working paper, Eindhoven University of Technology (2014)
17. Van der Heijden, M.C., Alvarez, E.M., Schutten, J.M.J.: Inventory reduction in spare parts networks by selective thoughput time reduction. Int. J. Prod. Econ. **143**, 509–517 (2013)

Answers to Selected Problems

Problem 2.1

(c) $\mathbf{S} = (3,2,1)$, $EBO(\mathbf{S}) = 0.1205$, $C(\mathbf{S}) = 8,300$ Euros

Problem 3.2

(c) $(S_A, S_B) = (3,3)$, $\hat{C}(S_A, S_B) = 892.7$ Euros/month, $W(S_A, S_B) = 0.1548$ days per demand

(d) Total average costs: 2,138 Euros/month, Increase: 388 Euros/month

Problem 4.2
$\mathbf{h}^* = (0,1)$, 644.26 Euros/month

Problem 4.3

(b) (0,0) is optimal for $0.1304 \leq \alpha < 1$, (0,1) is optimal for $0.0569 \leq \alpha \leq 0.1304$, (0,2) is optimal for $0 < \alpha \leq 0.0569$

Problem 5.1

(b) $\pi(0,0) = \frac{2}{37}$, $\pi(0,1) = \frac{5}{37}$, $\pi(1,0) = \frac{5}{37}$, $\pi(1,1) = \frac{25}{37}$

Problem 5.2

(b) $\pi(0,0) = 0.0881$, $\pi(0,1) = 0.1444$, $\pi(1,0) = 0.1976$, $\pi(1,1) = 0.5699$

G.J. van Houtum, B. Kranenburg, *Spare Parts Inventory Control under System Availability Constraints*, International Series in Operations Research & Management Science 227, DOI 10.1007/978-1-4899-7609-3

Problem 5.5

(b) $\beta = \frac{2+4m+3m^2}{2+6m+9m^2+9m^3}$, $\theta = \frac{9m^3}{2+6m+9m^2+9m^3}$, $A = \frac{2m+6m^2}{2+6m+9m^2+9m^3}$

Problem 6.1

(a) $\hat{S} = 4$, $S_{\text{tot}}^{(1)} = 24$, Corresponding mean waiting time: 0.0022 years
(b) 0.0004 years
(c) 7; 13; 11

Problem 7.1

$W_{1,0}(\mathbf{S}) = 0.0203$ months

Problem 8.2

(a) $L^1 = 9$
(b) The probabilities $P_j^1(\mathbf{U}_3)$ for $j = 0,\ldots,6$ are as follows: 0.11765, 0.10381, 0.09159, 0.08082, 0.07131, 0.06292, 0.05552
(c) The probabilities $P_j^2(\mathbf{U}_3)$ for $j = 0,1,2$ are as follows: 0.4, 0.24, 0.144

Problem 8.3

(c) 0.1322

Index

G.J. van Houtum, B. Kranenburg, *Spare Parts Inventory Control under System Availability Constraints*, International Series in Operations Research & Management Science 227, DOI 10.1007/978-1-4899-7609-3

MIX
Papier aus verantwortungsvollen Quellen
Paper from responsible sources
FSC® C105338

If you have any concerns about our products,
you can contact us on
ProductSafety@springernature.com

In case Publisher is established outside the EU,
the EU authorized representative is:
Springer Nature Customer Service Center GmbH
Europaplatz 3, 69115 Heidelberg, Germany

Printed by Libri Plureos GmbH
in Hamburg, Germany